ChromeOS System Administrator's Guide

Implement, manage, and optimize ChromeOS features effectively

Dr. Willie Sanders, Jr.

BIRMINGHAM—MUMBAI

ChromeOS System Administrator's Guide

Associate Group Product Manager: Mohd Riyan Khan
Publishing Product Manager: Prachi Sawant
Senior Content Development Editor: Adrija Mitra
Technical Editor: Arjun Varma
Copy Editor: Safis Editing
Project Coordinator: Sean Lobo
Proofreader: Safis Editing
Indexer: Subalakshmi Govindhan
Production Designer: Shyam Sundar Korumilli
Marketing Coordinator: Marylou De Mello

First published: February 2023

Production reference: 1120123

Published by Packt Publishing Ltd.
Livery Place
35 Livery Street
Birmingham
B3 2PB, UK.

978-1-80324-105-0

www.packtpub.com

Completing this book was truly a group effort. Without the encouragement of God, my family, and the team at Packt Publishing, finishing this project would not have been possible. Thank you all for enduring the late-night writing sessions and working around my crazy schedule to ensure deadlines were met. Most of all, thank you for believing in me enough to support this endeavor.

– Willie Sanders, Jr.

Contributors

About the author

Dr. Willie Sanders, Jr. is a veteran IT pro and professor of IT at Towson University (Towson, Maryland), where he motivates his students to succeed in IT. He's held senior technical roles with the US **Internal Revenue Service (IRS)** and at Pearson Education, where he worked as a senior systems administrator on their global IT team. Dr. Sanders is a **subject matter expert (SME)** who is heavily relied upon for his technical expertise and communication skills. Willie is also the founder of the nonprofit organization **Pass IT On (PITO)**, which provides free and low-cost technical training to disadvantaged youths and adults. Since 2015, PITO has impacted thousands of lives in Baltimore City, Maryland, and the surrounding region.

About the reviewers

Abhay Mukund Dhanki has been a technical specialist in the IT industry for more than 14 years, working with operating systems, servers, and networking. He has over 6 years of experience in cloud computing, SaaS, PaaS, and IaaS, involving the fundamentals of designing, building, and deploying productivity and collaboration solutions that meet the regional compliance and regulatory requirements of different organizations. He completed a post-graduate diploma in IT project management at Welingkar Institute of Management Development Research (Mumbai (Bombay), India) and also holds professional certifications in Linux, Networking (CCNA), Microsoft (MCITP), and Google (GCP).

I would like to thank my family, friends, and colleagues. It's rare to come across people who are so dedicated and trustworthy, and your efforts have not gone unnoticed. Working in this field would not have been possible without the supportive people that I have encountered over several years of learning. Thank you to all of the trailblazers who make this field an exciting one to work in every day. I am grateful for everything you do!

Abhishek Bhardwaj is a systems software engineer with experience in containers, virtualization, and security. He has contributed to the Chrome, Android, and ChromeOS code bases in C, C++, Java, and Rust. After obtaining his master's degree in electrical and computer engineering from Carnegie Mellon University, he worked for Microsoft and Google.

Table of Contents

Preface   xi

Part 1: Working with ChromeOS

1

ChromeOS Basics   3

Technical requirements	3	The Device menu	13
ChromeOS versus the competition	4	The Personalization menu	23
Cloud-focused	4	**OS updates, upgrades, and**	
Ease of use	4	**ChromeOS Flex**	**27**
Priced to move	4	Importance of ChromeOS updates	27
Getting up and running on ChromeOS	5	Upgrade time	27
Personalizing ChromeOS	7	Hardware flexibility with ChromeOS Flex	28
The desktop and its tools	7	Summary	28

2

Getting Connected   29

Technical requirements	29	Bluetooth and connected devices	39
Ethernet and Wi-Fi internet access	30	Connecting with Bluetooth	40
Connecting with Ethernet	30	Connecting an external display	41
Connecting with Wi-Fi	31	Connecting with Chromecast or smart TV	41
Virtual private network (VPN)	32	Casting and syncing	42
Mobile data connectivity	34	Casting from Google Chrome	42
Built-in mobile data	35	Casting from your desktop	43
Mobile data connection using a dongle	36	Syncing with Android devices	44
Mobile data connection using an Android device	37	Summary	46

3

Exploring Google Apps   47

Technical requirements	**47**	Sheets	58
Managing default apps and		Slides	58
notifications	**47**	Keep	59
Default apps	48	Forms	59
Notifications	51	Jamboard	60
		Freemium versus premium plans	61
Google Workspace applications	**53**	Premium tiers	62
Gmail	54		
Drive	55	**Accessing Chrome browser-based apps**	**63**
Meet	55	**The Chrome Web Store**	**64**
Calendar	56	**Google Play Store**	**65**
Chat	57	**Summary**	**67**
Docs	57		

Part 2: Security and Troubleshooting

4

ChromeOS Security   71

Technical requirements	**71**	2FA	80
User account management	**72**	**System updates, sandboxing, and**	
Secondary user accounts	72	**verified boot**	**82**
Sync	74	System updates	82
Other Google services	76	Sandboxing	85
Guest browsing	77	Verified boot	85
Advanced user management	78		
		Parental controls	**86**
Screen locks and 2FA	**79**	**Chrome browser-based settings**	**89**
Screen locks	79	**Summary**	**91**

5

Recovering from Disasters   93

Technical requirements   93   Backing up to Google Drive   99
Hardware and cloud-based data   Backing up to external storage   101
encryption   94
                                                      OS and app recovery   102
What is encryption?   94   When should you recover ChromeOS?   102
ChromeOS hardware-based encryption   95   Recovery mode   102
ChromeOS software-based encryption   97   USB-based OS recovery method   103

Data backup strategies for ChromeOS   98   Summary   107
Backup basics   98

6

Troubleshooting 101   109

Technical requirements   109   No network connectivity   118
Crashing and freezing   110   Missing OS   121
Slow system performance   111   Summary   122
Constantly refreshing tabs   116

Part 3: Advanced Administration

7

The Linux Development Environment   125

Technical requirements   125   Linux permissions   132
Enabling the Linux development   Linux backup/restore   134
environment   126
                                                      Managing Linux storage   136
What is the Linux development environment? 126
Turning on the LDE   126   Enabling Android Debug Bridge   138
Linux features   128   Implementing port forwarding   141
                                                      Summary   144
Managing device permissions and
Linux backup/restore   132

8

Working with the Chrome Shell (Crosh)   145

Technical requirements	145	battery_firmware info	152
Accessing Crosh	146	free	152
Looking at the essential Crosh		meminfo	152
commands for systems administration	146	memeory_test	153
		Ssorage_test	154
help	147	network_diag	155
help_advanced	147	tracepath	155
uptime	147	route	156
top	148	rollback	157
enroll_status	149	vmc	157
ipaddrs	149	exit	157
ping	150		
modem	150	**Shell scripting**	**157**
set_time	151	**Summary**	**160**
battery_test	151		

9

Google Workspace Admin Console   161

Technical requirements	161	Reporting	178
An overview of Google Workspace	162	Apps	185
The major features of the Admin		Security	190
console	170	Storage	195
Accessing the Admin console	171	Rules	196
Account	174	**Migrating to Google Workspace**	**197**
Billing	177	**Summary**	**198**

10

Centralized Administration of OUs, Users, Groups, and Devices   199

Technical requirements	200	Exploring the Directory and Devices	
		submenus in the Admin console	200
		Directory	200

Devices   205

Getting organized with OUs   212

**Exploring advanced user account
management**   214

Adding or editing multiple user accounts in
your Google Workspace   218

Managing the custom attributes for user
accounts   220

**Understanding groups and target
audiences**   222

**Exploring advanced device
management**   226

**Exploring advanced application
management**   228

Summary   230

Index   231

Other Books You May Enjoy   240

Preface

Google's ChromeOS provides a great platform for technicians, systems administrators, developers, and casual users alike. ChromeOS provides a seemingly simplistic architecture that is easy enough for a novice user to begin working with immediately. However, beneath the surface, this operating system boasts a plethora of powerful tools, able to rival any other OS on the market, so learning how to harness the full potential of the OS is critical for technical workers and users. What sysadmins need is a guide to the OS's unique features and capabilities... and here it is: *ChromeOS System Administrator's Guide*.

This book explains ChromeOS in terms of the following:

- Its unique architecture
- Built-in tools for performing essential tasks
- Data management capabilities
- Cloud-first application implementation functionality

As you build your foundational knowledge of the OS, you'll be exposed to higher-level concepts such as the following:

- Security
- The command line
- Enterprise management

By the end of this book, you will be perfectly placed to perform a range of system administration tasks within ChromeOS without requiring an alternative operating system, thereby broadening your options as a technician, sysadmin, developer, or engineer.

Who this book is for

This book is for system administrators, developers, and engineers who want to explore the ChromeOS architecture and its administration tools and techniques. Basic knowledge of system administration is required.

What this book covers

Chapter 1, ChromeOS Basics, takes a closer look at the features and functions of ChromeOS. We will learn how to use ChromeOS's built-in tools to perform essential tasks including managing user accounts, working with data, and launching applications.

Chapter 2, Getting Connected, shows the multiple ways offered by ChromeOS to connect your device to the world around you. This chapter introduces the operating system's approach to connecting to the internet via wired and wireless means, along with connecting to other devices.

Chapter 3, Exploring Google Apps, examines the host of cloud-based applications offered by Google that can be used natively by ChromeOS. Additionally, because of its Google Play Store integration, ChromeOS can also access thousands of additional apps. This chapter explains the uses of the most common applications you'll find on ChromeOS, how to access them, and how to acquire additional applications.

Chapter 4, ChromeOS Security, discusses how, although ChromeOS is a streamlined and secure OS platform, it still has the potential to be compromised if the right protections aren't in place. In this chapter, we discuss the common ways you can harden your ChromeOS setup against cyber attacks.

Chapter 5, Recovering from Disasters, considers that while hardware and software are extremely important aspects of a computer system, the data that they produce and manage is of even greater value. ChromeOS has several built-in features that users can leverage to ensure the applications and the data they create remain safe and, in the worst-case scenario, are easy to recover.

Chapter 6, Troubleshooting 101, deals with the fact that although ChromeOS is among the safest, most secure, and easiest-to-use operating systems on the market, it is still susceptible to failures from time to time. This chapter explores some common issues you may encounter on systems running ChromeOS and how to fix them.

Chapter 7, The Linux Development Environment, acknowledges that for software developers, setting up an environment for writing code and developing apps can sometimes be a hassle. Thankfully, ChromeOS simplifies this task by providing a Linux virtual machine. In this chapter, we explore the Linux Development Environment and its key features.

Chapter 8, Working with the Chrome Shell (Crosh), highlights how, in addition to ChromeOS's graphical user interface, advanced users can also leverage the Chrome Shell (crosh). This powerful application gives ChromeOS system administrators access to power tools and utilities not available in the GUI.

Chapter 9, Google Workspace Admin Console, notes that with the popularity of Chromebooks on the rise, more and more organizations are beginning to adopt them as their computers of choice. But even though the ChromeOS has powerful local administration tools, it still needs help when introduced into corporate settings where centralized management is key. In this chapter, you'll learn how Google Workspace and its Admin console help ChromeOS meet the needs of enterprise.

Chapter 10, Centralized Administration of OUs, Users, Groups, and Devices, considers how the centralized management of resources is the hallmark of enterprise system administration. In this chapter, we learn how locally created users, groups, and ChromeOS devices themselves can be managed from the Google Workspace Admin console.

To get the most out of this book

Having an understanding of basic client-side computing and administration concepts will be helpful as you work through the concepts in this book. That being said, the book is written to help any reader upskill to becoming a ChromeOS systems administrator, so no previous systems administration experience is necessary.

Software/hardware covered in the book	Operating system requirements
Chromebook and Chromebox	ChromeOS or ChromeOS Flex
Google Workspace	
Chrome web browser	

Download the color images

We also provide a PDF file that has color images of the screenshots and diagrams used in this book. You can download it here: `https://packt.link/HOScT`.

Conventions used

There are a number of text conventions used throughout this book.

Bold: Indicates a new term, an important word, or words that you see onscreen. For instance, words in menus or dialog boxes appear in **bold**. Here is an example: "Navigate to the **Settings** menu and select **Accounts**."

Italics: Indicates references to another chapter, another section in the same chapter, or a particular image in a chapter. This style is also used to highlight the keyboard keys. Here is an example: "ChromeOS will then complete the process by confirming that your accounts are linked and providing you with a review screen that lists the apps that are currently installed on the device along with their maturity ratings, as seen in *Figure 4.15*."

> **Tips or important notes**
> Appear like this.

Get in touch

Feedback from our readers is always welcome.

General feedback: If you have questions about any aspect of this book, email us at `customercare@packtpub.com` and mention the book title in the subject of your message.

Errata: Although we have taken every care to ensure the accuracy of our content, mistakes do happen. If you have found a mistake in this book, we would be grateful if you would report this to us. Please visit `www.packtpub.com/support/errata` and fill in the form.

Piracy: If you come across any illegal copies of our works in any form on the internet, we would be grateful if you would provide us with the location address or website name. Please contact us at `copyright@packt.com` with a link to the material.

If you are interested in becoming an author: If there is a topic that you have expertise in and you are interested in either writing or contributing to a book, please visit `authors.packtpub.com`.

Share Your Thoughts

Once you've read *ChromeOS System Administrator's Guide*, we'd love to hear your thoughts! Scan the QR code below to go straight to the Amazon review page for this book and share your feedback.

`https://packt.link/r/1803241055`

Your review is important to us and the tech community and will help us make sure we're delivering excellent quality content.

Download a free PDF copy of this book

Thanks for purchasing this book!

Do you like to read on the go but are unable to carry your print books everywhere? Is your eBook purchase not compatible with the device of your choice?

Don't worry, now with every Packt book you get a DRM-free PDF version of that book at no cost.

Read anywhere, any place, on any device. Search, copy, and paste code from your favorite technical books directly into your application.

The perks don't stop there, you can get exclusive access to discounts, newsletters, and great free content in your inbox daily

Follow these simple steps to get the benefits:

1. Scan the QR code or visit the link below

https://packt.link/free-ebook/9781803241050

2. Submit your proof of purchase
3. That's it! We'll send your free PDF and other benefits to your email directly

Part 1: Working with ChromeOS

In this part, we will understand the basics of ChromeOS as an operating system, its architecture, and the key features that set it apart as a cloud-first operating system.

This part comprises the following chapters:

- *Chapter 1, ChromeOS Basics*
- *Chapter 2, Getting Connected*
- *Chapter 3, Exploring Google Apps*

1

ChromeOS Basics

ChromeOS is a ground-breaking, **cloud-first** (designed to take advantage of cloud services) **operating system (OS)** developed by Google. Its affordability, unique design, low hardware requirements, and ease of use have made it a popular choice with consumers of all types. From bedrooms to boardrooms, you'll find ChromeOS being used. So, as an administrator, it is important that you understand how to best leverage ChromeOS's power and flexibility.

In this chapter, we'll take a look at the core features and functions of ChromeOS. You'll learn how to use ChromeOS's built-in tools to perform essential tasks such as managing user accounts, working with data, and launching applications. By the end of the chapter, you'll know many of the ins and outs of working with ChromeOS and why it's so different from other OSs on the market.

Ready to get started? *We thought so!*

In this chapter, we will be covering the following topics:

- ChromeOS versus the competition
- Getting up and running on ChromeOS
- Personalizing ChromeOS
- OS updates, upgrades, and ChromeOS Flex

Technical requirements

In order to get the most out of your introduction to the fundamental features of ChromeOS, you'll need access to the OS. You have two options:

- Use Chromebook or Chromebox hardware, which comes with ChromeOS natively installed
- Install **ChromeOS Flex** (which we will discuss later in the chapter) onto a PC or Mac

ChromeOS versus the competition

An OS is the most important piece of software you can install on your computer. Without it, we can't issue commands to our hardware, organize files, or even log in to a system. It is essential! Therefore, it's no wonder why there are several major OS platforms on the market.

With major competitors such as Microsoft and Apple, you may ask, *"What does ChromeOS have to offer that they don't?"* In this section, we will explore some of the things that help ChromeOS stand out from the crowd.

Cloud-focused

Most computer OSs are designed with a focus on localized computing and storage. This means that these OSs rely heavily on the hardware components installed in your computer to perform data processing and storage. This design is fine if you rely on locally installed applications, but is not as necessary if you primarily use cloud-based applications.

As the world's first commercially successful, cloud-first OS, ChromeOS shifts the burden of processing and storage from your device to the cloud. It does this by building on the foundational technologies introduced in Google's popular Chrome web browser. In fact, as you'll see later in the chapter, the Chrome browser is the central component of ChromeOS. This cloud-based approach allows hardware manufacturers to focus on developing lower-power, terminal-like, computer devices that are built to take full advantage of all of the benefits that moving to the cloud has to offer.

Ease of use

Managing and maintaining an OS can be challenging. Traditional OSs can be very complex because they are responsible for managing local hardware, applications, processing, storage, data I/O, and so much more. This complexity trickles down to you, the administrator, who now has to learn how to navigate and configure the hundreds of settings each OS provides. This can be a daunting task and it's why there are entire courses and books dedicated to the subject.

ChromeOS takes a different approach. By shifting many of the duties of a traditional OS to the cloud, what you are left with is a sleek, simplistic, bowser-centric OS. The reduced number of configurations and settings make learning how to manage the OS a very straightforward process, empowering users of all skill levels to control the management of their systems.

Priced to move

As previously mentioned, traditional OS designs rely on the use of directly attached hardware. However, there is a major drawback to this approach. In order to guarantee high levels of performance for the OS running these hardware platforms, vendors have used higher-end hardware components. The inclusion of these higher-end components in turn drives up the overall cost of the computer device and that cost gets passed along to you, the consumer.

Google has been able to use the cloud-first design of ChromeOS to create a niche in the computer hardware market. ChromeOS's streamlined design and heavy reliance on the cloud for computing power and storage have resulted in its ability to run on less expensive hardware. As a result, hardware manufacturers can create **Chromebook** (laptops designed with ChromeOS as their native OS) and **Chromebox** (desktops designed with ChromeOS as their native OS) computers at a fraction of the cost of other computer devices. These savings are then passed along to consumers resulting in inexpensive but reliable computers.

Now that you know a little about how ChromeOS compares to the competition, let's look at how to get started using the platform.

Getting up and running on ChromeOS

ChromeOS is the native OS of Chromebox desktops and Chromebook laptops. This means that out of the box you have access to the OS. Much like its mobile device counterpart, Google's *Android OS*, ChromeOS requires the creation of or the use of an existing Google Account in order to access the full functionality of the system.

The first time you power on the system, you will be greeted with the **Who's using this Chromebook?** screen, as seen in the following figure:

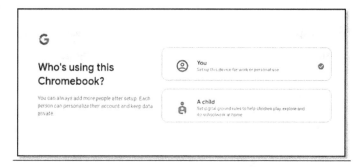

Figure 1.1 – The Who's using this Chromebook? setup screen

The primary options on this screen provide you with links to set up your personal user account or to set up a restricted account for a child. However, hidden below these options, you'll find links that will allow you to browse as a guest or perform enterprise enrollment of your device. Browsing as a guest allows you to bypass entering your Google account information by providing you with a limited computing environment. This is ideal for someone who just needs to use the internet on a Chrome device that doesn't belong to them. Enterprise enrollment is a feature used by organizations that want to have centralized management over multiple devices running ChromeOS. We will discuss enterprise enrollment in more depth in *Chapter 9, Google Workspace Admin Console*.

Assuming you will be the primary user of the Chrome device, perform the following steps to set up your account:

1. On the **Who's using this Chromebook?** Screen, select the **You** button, then press **Next**.

2. You'll be presented with a **Sign into your Chromebook** screen. At this point, you will choose to do one of the following options:

    -   Enter the email address or phone number of your existing Google Account.

    -   Select the **More Options** link, which will allow you to create a new account or perform enterprise enrollment. Let's choose to create a new account.

3. On the **Create a Google Account** screen, enter your first name and last name and press **Next**.

4. You are then prompted to enter some basic information, including your birthday and gender. Enter the required information and press **Next**.

5. On the **Choose your Gmail address** screen, you will be given several suggestions for potential Gmail addresses. You can either choose one of these addresses for your account or create your own unique email address. After you have chosen or created an email address for your account, press **Next**.

6. Now you'll be prompted to create a password. In order to ensure that your password isn't easily guessed, be sure to implement **password complexity** by using a mixture of upper and lower-case letters, numbers, and special symbols. Once you have a strong password in place, press **Next**.

7. (Optional) You can include a phone number on the **Add phone number?** screen. Adding a phone number provides you with the following benefits:

    -   *Password recovery* if you forget your ChromeOS login

    -   The ability to receive video calls and messages

    -   Connectivity to other Google services

8. On the **Review your account info** screen, you have one final chance to verify your Google account's email address and username. If these settings are correct, then you can click **Next**.

9. Google's **Privacy and Terms** policy is then shown. Once you've read and agreed to it, you can click the **I agree** button.

10. (Optional) The **Sync your Chromebook/Chromebox** screen gives the opportunity to sync ChromeOS with your Chrome browser settings from other devices. You can select **Accept and continue** to perform the sync:

Note

You can bypass this section by selecting the **Review sync options following setup** checkbox option.

---

11. (Optional) By pressing the **More** button on the **Google Play apps and services** screen, you'll be provided with the terms and conditions for using the Google Play store on your ChromeOS device. This feature will be discussed in more detail in *Chapter 3, Exploring Google Apps*. You will also have the option to enable the **Back up to Google Drive** feature, which we'll dive into in *Chapter 5, Recovering from Disasters*. Once you've viewed the terms and made your selection on backups, click the **Accept** button to continue.

> **Note**
>
> You can bypass this section by selecting the **Review Google Play options following setup** checkbox option.

12. (Optional) On the **Your Google Assistant is ready to help** screen, you get the option of enabling **Google Assistant** to help you with managing ChromeOS and its apps via voice commands. Select the **No thanks** button to leave this option disabled or the **I agree** button to activate it:

    We will discuss Google Assistant in more detail later in this chapter

13. Finally, you'll arrive at the **You are all set!** screen. Here you'll have the option of signing up to receive email communications from Google. Once you've made that choice, you can click the **Get started** button to fully log in to the system and view your ChromeOS desktop.

Now that we have the initial setup on our ChromeOS device taken care of, let's look at how we can personalize the OS to make it our own.

Personalizing ChromeOS

Although ChromeOS is designed with simplicity in mind, it still provides users with tons of opportunities to modify its look and feel. Some of the available configurations are purely cosmetic but many of them can have a profound impact on user experience. As a systems administrator, it will be important for you to understand how to locate and adjust these settings to ensure you and your users have the ideal computing experience.

The desktop and its tools

Like other OS, ChromeOS provides users with a **desktop** workspace once they've successfully logged into the system. However, this desktop differs from most OS desktops because it does not allow file, folder, or applications shortcut icons to be stored on it. Instead, it provides you with adjustable wallpaper graphics and a centralized location to access primary tools for organizing and navigating in ChromeOS: **Shelf**, **App Launcher** and **Status Tray**, as shown here:

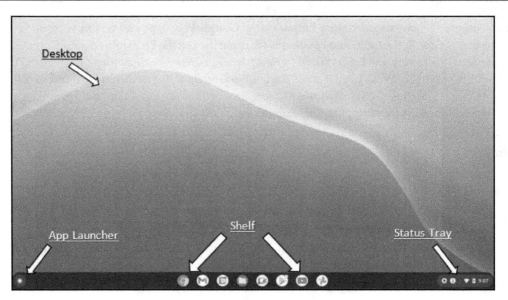

Figure 1.2 – The ChromeOS desktop

Let's look at each tool in detail in the following sections.

The shelf

The **shelf** acts as a storage location for tools and apps in ChromeOS. The shelf is home to **App Launcher** and the **status tray**. It also displays shortcuts for both pinned and running apps. Apps can be pinned or unpinned from the shelf by using the following steps:

- Pinning apps:

  I.   Locate the app that you want to pin using App Launcher.

  II.   Click and hold the chosen app.

  III. Drag the app's icon to the shelf.

- Unpinning apps:

  I.   Locate the app which you want to unpin on the shelf.

  II.   Click and hold the chosen app.

  III. Drag the app's icon off of the shelf.

Pro-tip

By default, the shelf is located at the bottom of your screen on the desktop, but it doesn't have to be. To adjust the position of the shelf and to make it automatically hide when not in use, do the following:

1. Right-click (using the mouse) or tap with two fingers (using the touchpad) to open the **Quick Settings** menu.

2. Select the **Shelf position** menu option and then choose **left**, **bottom**, or **right** to move the shelf to its new location.

3. Or, select the **Autohide Shelf** menu option to allow the shelf to collapse from view when another screen opens.

Figure 1.3 illustrates how you can rearrange the shelf to suit your needs.

Figure 1.3 – The shelf moved to the right position

App Launcher

App Launcher is your primary tool for navigating ChromeOS and the internet. It provides you with search capabilities and shortcuts to popular tools and apps. You also use App Launcher to access the full selection of utilities and apps that ChromeOS has to offer. Simply click on App Launcher's icon and expand it to full screen to see everything you have installed on your ChromeOS device. Here's what that looks like:

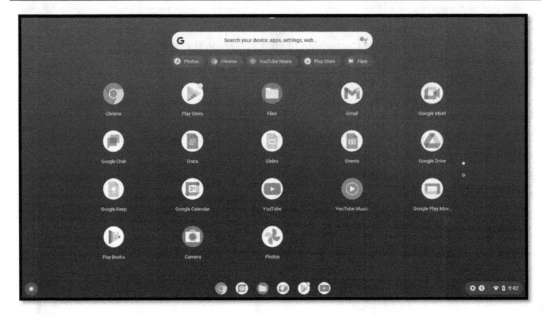

Figure 1.4 – App Launcher in normal full-screen view

> **Pro-tip**
>
> If you prefer more of a Microsoft Windows-style navigational experience, you can adjust App Launcher by activating one of ChromeOS's many *experimental* features. To access this enhancement, do the following:
>
> 1. In your Chrome browser, search for `chrome://flags/#productivity-launcher`.
>
> 2. Locate the **Productivity experiment: App Launcher** option.
>
> 3. Change its configuration setting from **Default** to **Enabled**.
>
> 4. Press the **Restart** button to reboot ChromeOS and apply the change.

Figure 1.5 shows how the **Productivity experiment: App Launcher** flag will appear in your Chrome browser.

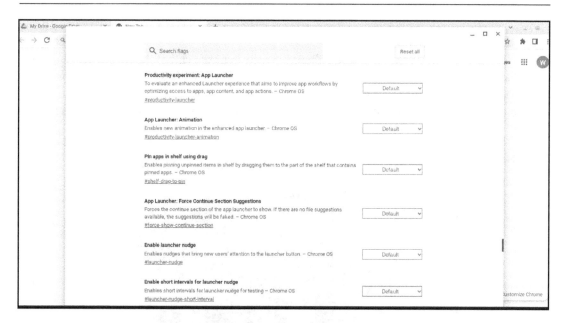

Figure 1.5 – ChromeOS experimental features

Once the flag is applied, your App Launcher screen takes on a *Windows-style* appearance, as seen in *Figure 1.6*.

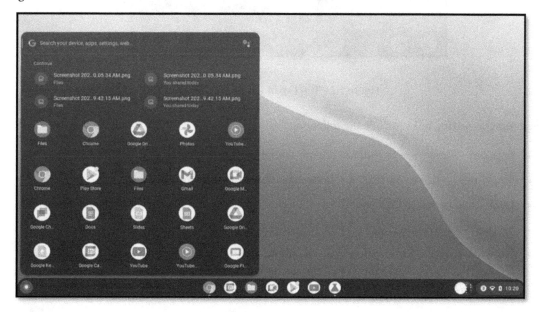

Figure 1.6 – ChromeOS with Windows-style App Launcher enabled

The status tray

Now that we have explored App Launcher, let's move to the opposite end of the shelf where we will find the status tray. In its default state, the status tray displays indicator icons for your system's battery and network. It's also where you'll see the time and settings. However, when you click on this tool to expand it, you'll be presented with a myriad of additional settings as seen in the following figure:

Figure 1.7 – The expanded status tray

The status tray setting links include the following:

- **Sign out** – allows you to sign out of your Google account.
- **Shutdown** – allows you to do a software shutdown of your ChromeOS device.
- **Lock** – allows you to lock your devices, forcing a password to be entered to regain access.
- **Settings** – allows quick access to the full settings menu discussed later in this chapter.
- **Network Connections** – allows you to choose a wired or wireless network connection. For wireless networks, it also shows you the strength of your wireless signal.

    We will discuss networking with ChromeOS in more detail in *Chapter 2, Getting Connected*.

- **Bluetooth** – enables or disables Bluetooth functionality. It also allows you to scan, pair, and connect to Bluetooth-capable devices (such as Bluetooth headphones and speakers).

- **Notifications** – allows you to configure which apps can give you pop-up notifications. You can also turn on the **Do not disturb** option, which blocks all notifications.

- **Screen Capture** – allows you to take still-frame pictures of your display or record video of your on-screen actions. Both still-frame pictures and videos can then be edited and saved to your device or the cloud.

- **Nearby Visibility** – allow your device to be seen on your network by other devices running ChromeOS or Android OS for the purpose of sharing files.

  We will discuss connecting to other devices with ChromeOS in more detail in *Chapter 2, Getting Connected.*

- **Night Light** – allows you to manually or automatically dim your display to make it easier to read in low light.

- **Cast** – allows you to (wirelessly) share your full display or a specific Chrome browser tab with a wireless-capable monitor or television. You can also connect to devices that are using Google's Chromecast hardware.

  Chromecast is a plug-in hardware component that gives normal televisions smart TV functionality.

- **Volume/Audio Settings** – allows you to adjust the speaker volume for your system using a slider bar, control speaker, and microphone settings using its advanced features.

  Advanced speaker and microphone settings include the ability to choose which speaker or microphone you would like to use. This is useful in situations where you have multiple speakers or mics connected to the ChromeOS device.

- **Brightness** – allows you to control how bright or dim your display is.

  Controlling the screen brightness is a good way to make it easier to read your screen in different lighting situations. It's also useful when you need to conserve battery power (dimming your display can extend your battery power considerably).

Now that we've explored the desktop and its tools, let's take a closer look at how we can manage our ChromeOS device using tools in the **Device** menu.

The Device menu

As you continue to customize ChromeOS, you will need to explore the **Settings** screen. This area provides you with a centralized location for managing all of the configuration options ChromeOS has available. The configurable options are grouped together by searchable categories, making finding what you're looking for a very straightforward process.

One of the **Settings** screen's categories that you will find useful during the setup of your ChromeOS device is the **Device** category, as shown in the following screenshot:

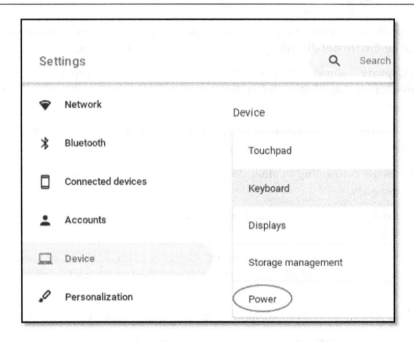

Figure 1.8 – The Device category of the Settings screen

The **Device** category provides you with the following configuration options:

- Mouse/touchpad
- Keyboard
- Displays
- Storage management
- Power

Let's explore each to see how the administrator can use them to improve the user experience.

Mouse and touchpad

Under **Device → Mouse** and **Touchpad**, you will find settings to adjust the functionality of your mouse or touchpad device, as shown in *Figure 1.9*. If you are on a ChromeOS device that only has one of the two input devices, you'll only see options for adjusting that device's settings (for example, a Chromebook with no external mouse will only show touchpad settings).

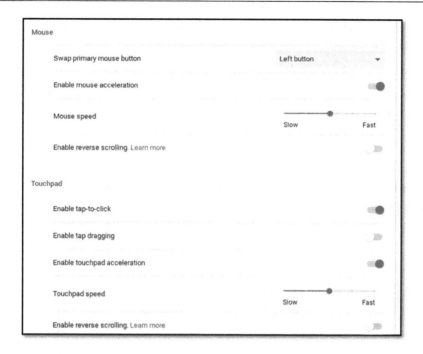

Figure 1.9 – The Mouse and Touchpad settings screen

The **Mouse** settings include the following:

- **Swap primary mouse button** – allows you to switch the functionality of the left and right mouse buttons.

- **Enable mouse acceleration** – allows you to adjust how fast or slow the mouse moves your pointer across the screen.

  The mouse speed slider is only active when **Enable mouse acceleration** has been enabled.

- **Enable reverse scrolling** – makes upward scrolling action move pages downward and downward scrolling action move pages upward.

The **Touchpad** settings include the following:

- **Enable tap-to-click** – allows you to enable or disable the ability to tap the touchpad to select an object instead of clicking on it.

- **Enable tap dragging** – allows you to enable or disable the ability to double-tap an object and drag it around the screen.

  The alternative method for dragging objects is to use the touchpad to click and hold an object with one finger while using another finger to drag the object.

- **Enable touchpad acceleration** – similar to mouse acceleration, this setting allows you to control how fast or slow the touchpad moves your pointer across the screen.

  The touchpad speed slider is only active when **Enable touchpad acceleration** has been enabled.

- **Enable reverse scrolling** – this setting functions the same with the touchpad as it does with a mouse. It makes the upward scrolling action move pages downward and downward scrolling action move pages upward.

Keyboard

Under **Device** → **Keyboard**, you will find several settings that allow you to customize how certain keys on a ChromeOS device's keyboard function, as shown in *Figure 1.10*. You can also adjust the sensitivity of the keyboard to accommodate users who may have physical conditions that affect their typing abilities.

Figure 1.10 – The Keyboard settings screen

The **Keyboard** settings include the following:

- The ability to change the keys used to represent the **Search, Ctrl, Alt, Escape,** and **Backspace** functions.

- **Treat top-row as function keys** – allows you to enable or disable the ability to hold the **Search** key in order to change the behavior of the top-row keys.

- **Enable auto-repeat** – allows you to enable to disable the ability to adjust how long your ChromeOS device will wait after you press a key before it begins to repeat. You can also adjust how fast or slow a key repeats:

  - When enabled, you can adjust the **Delay before repeat** and **Repeat rate** settings using their sliders

- **Change input setting** – allows you to access a sub-menu with advanced settings that enable you to do the following:

  - Show an input options shortcut on the shelf

  - Add additional keyboard input languages

  - Customize the spelling and grammar check dictionary, as shown in the following figure:

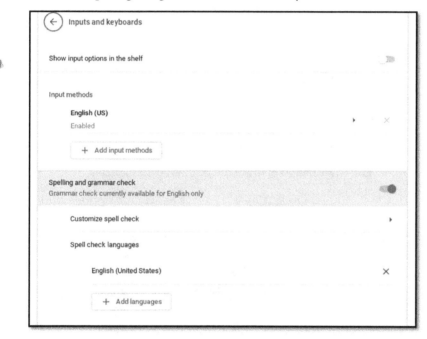

Figure 1.11 – The Inputs and keyboards settings screen

Pro-tip

The mouse and touchpad are great tools to use when navigating the many features and functions of ChromeOS. However, if you really want to increase your productivity, then you'll need to master the many keyboard shortcuts available to you in ChromeOS.

To see a full listing of these handy shortcuts, select the **View keyboard shortcuts** link on the **Devices** →**Keyboard** settings screen, and this is what you'll get:

Figure 1.12 – Sample of the ChromeOS keyboard shortcuts

Displays

Under **Device** →**Displays**, you will find settings that allow you to configure how your ChromeOS devices' built-in and connected displays operate. The settings for the built-in display are simple and limited but once an external display is connected, a host of additional configuration options become available for you to use as you tailor the monitor's settings to your viewing needs.

The **Built-in display** settings include the following:

- **Display size** – allows you to adjust how small or large items appear on the screen.

  This setting emulates making changes to a screen's resolution.

- **Orientation** – allows you to rotate the screen 0, 90, 180, or 270 degrees.

- **Night Light** – as mentioned earlier in the chapter, this feature adjusts the screen to make it easier to read in dim light. Once enabled, you also have the option of activating this feature on an automated schedule.

Here's what they look like on screen:

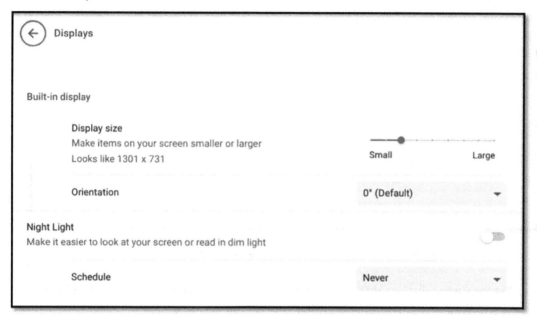

Figure 1.13 – The Built-in display settings

The external display settings include the following:

- **Arrangement** – allows you to determine the order of extended monitors by dragging them or using the arrow key to move them into your desired order. This affects how your pointer moves from screen to screen when your displays are in **Extended Mode**.

  The **Arrangement** setting also provides an option to duplicate the built-in display. You can enable this option by clicking the checkbox labeled **Mirror Built-in display** as shown in the following figure:

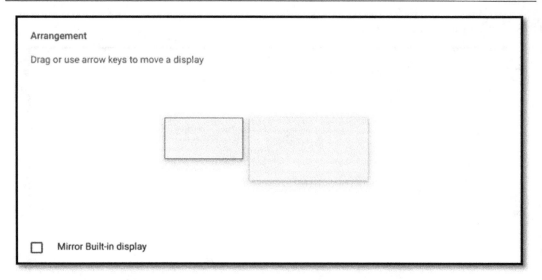

Figure 1.14 – The Arrangement settings for an external display

- **Screen** – when displays are in **Extended Mode**, this setting allows you to select which screen will be the primary display, and which will be the extended display.

- **Display size** – the same as the built-in display settings.

- **Resolution** – allows you to adjust the display resolution, which affects how large or small images and text appear on the screen.

- **Refresh Rate** – allows you to configure how often the image on your display is updated per second.

  Higher refresh rates cause the movement of images on your screen to appear smoother.

- **Orientation** – The same as the built-in display settings.

- **Overscan** – allows you to adjust the boundaries of your display to ensure that the edges of your desktop display are not cut off from view.

- **Night Light** –The same as the built-in display settings.

Figure 1.15 illustrates how each of the previously mentioned configuration options will appear on-screen:

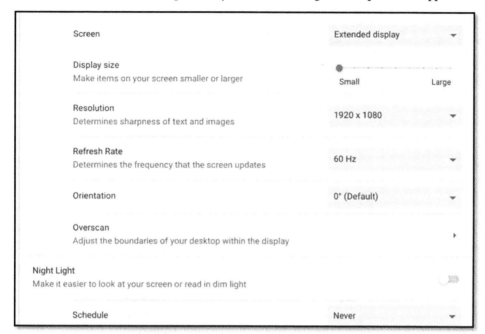

Figure 1.15 – The other external display settings

Storage management

Under **Device →Storage management**, you won't find many customizations, as seen in *Figure 1.16*. However, you will be provided with several important pieces of information regarding storage resource usage on your ChromeOS device. Additionally, you will be presented with several links to other areas of the OS and Chrome browser that will allow you to customize how your device's storage is used and maintained.

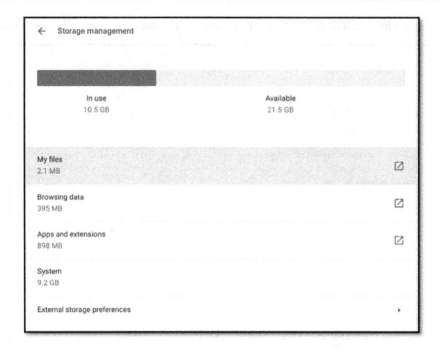

Figure 1.16 – The Storage management settings

The **Storage management** screen displays the following:

- The amount of used and available storage in your ChromeOS device, in bar chart form.
- A **My files** link, which redirects you to the ChromeOS file manager, known as **Files**.

  This is the default storage location for any files that you create or download to your ChromeOS device.
- A **Browsing data** link that redirects you to the Chrome browser's **Clear browsing data** screen.

  We will discuss managing browsing data in-depth in *Chapter 4, ChromeOS Security.*
- An **Apps and Extensions** link that redirects you to the **Device → Apps** section of the settings menu where you can see and manage the apps that are available to your ChromeOS device.

  We will discuss accessing and managing apps in more detail in *Chapter 3, Exploring Google Apps.*

Power

Under **Device → Power**, you will be able to access information about your ChromeOS device's current power usage, as seen in the following image. Also, you'll be provided with a few options for changing power-related settings.

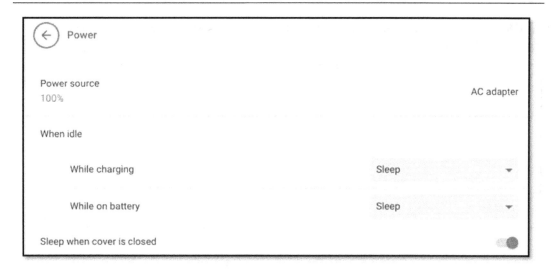

Figure 1.17 – The Power settings

The **Power** settings menu includes the following:

- **Power Source** – displays if your ChromeOS device is currently using the AC adapter or battery as a power source. When on battery power, you will also be shown the percentage of the remaining battery and an estimate of how many hours/minutes you have before the device needs to be recharged.

- **When idle** – allow you to configure what happens to your ChromeOS device when it is being actively used. On ChromeOS devices with batteries (such as Chromebooks), you'll be able to define different settings using the AC adapter versus using battery power:

  - **While Charging** – controls what happens to idle systems while they are being powered by the AC adapter; the options are **Sleep**, **Turn off display**, and **Keep display on**

  - **While on battery** – controls what happens to idle systems while they are being powered by battery; the options are **Sleep**, **Turn off display**, and **Keep display on**

- **Sleep when cover is closed** – provides enable/disable toggle switch that allows you to control if a ChromeOS device with a lid cover (such as a Chromebook) goes to sleep or stays active when the device's cover is closed.

The Personalization menu

Many of the customization settings addressed in the **Device** menu affect the functionality of ChromeOS and the device that it's installed on. But sometimes customizing your OS is more about aesthetics than functionality. This is why ChromeOS also includes several options for customizing the look and feel of the OS. These settings are grouped together under the **Settings → Personalization** menu, as you can see in the following figure:

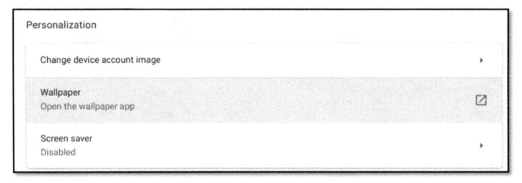

Figure 1.18 – The Personalization menu

In this section, we'll discuss some of the key configurations that you'll be able to alter to make ChromeOS your own.

Account image

Under **Personalization → Change device account image**, you're given the ability to change the picture that is associated with your account on the sign-in page, as you can see in *Figure 1.19*. This is a good way to visually distinguish between multiple accounts on your ChromeOS devices. It's also a cool way to show off your own unique style and personality.

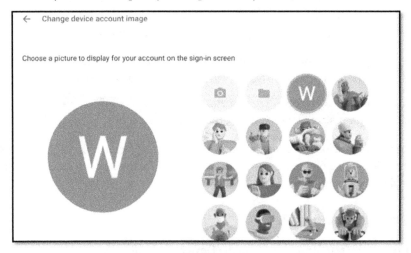

Figure 1.19 – The Change device account image options

Choose between the following options when selecting your account image:

- Keep the default, which is typically the initials of your user account name

- Choose from an array of ChromeOS-provided account images

- Take a picture with your webcam

- Select a stored image from **My files**

Wallpapers

Under **Personalization → Wallpaper**, you'll have access to a link that redirects you to the **Wallpaper** app, as seen in *Figure 1.20*. Configuring a wallpaper will change the image that appears on the desktop screen when you first sign in to ChromeOS.

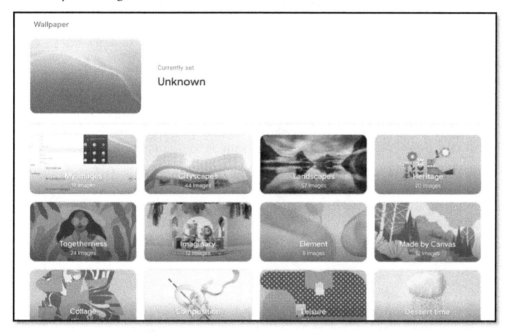

Figure 1.20 – The Wallpaper app

In the **Wallpaper** app, you are presented with the option of choosing from images that you have stored in **My Files** or collections of images made available by Google. Once you've chosen your image source, you also get the option to choose a single, static wallpaper or to have the wallpaper automatically change every day. Let's see what this page looks like:

Figure 1.21 – The change daily option for Wallpaper

Screen saver

Under **Personalization → Screen saver**, you'll be presented with a toggle switch to enable/disable the ChromeOS screen saver as you can see in the following image. The screen saver displays when the system is idle. It can be configured to show photos, the time, the weather, and media information. Enabling this setting will also allow your display to remain on while using an AC adapter.

Figure 1.22 – The Screen saver settings menu

The **Screen saver** menu options include the following:

- **Background** – allows you to configure the source of your screen saver images as **Google Photos** or the **Art gallery**
- **Weather** – allows you to select the temperature measurement used for the weather on the screen saver

Now that you've made it through this section, not only have you mastered configuring many of ChromeOS's core settings, but you've also learned how to give it your own personal touch and customized look and feel with the personalization menu.

In the next section, you will get a preview of how you can keep Chrome up to date with the latest and greatest features. We will also explore a new, breakthrough implementation of the OS; ChromeOS Flex.

OS updates, upgrades, and ChromeOS Flex

Now that you have ChromeOS up and running, let's talk about keeping it up and running! A large part of maintaining a healthy OS is ensuring that routine maintenance activities are performed. In this section, we will introduce two of the most important maintenance tasks that administrators can perform on their ChromeOS devices: **updates** and **upgrades**. We will also explore Google's solution for making ChromeOS more hardware independent with the release of ChromeOS Flex.

Importance of ChromeOS updates

Updates are an important task on ChromeOS. Administrators need to ensure that updates are being performed on a regular basis, so the latest and greatest tools and features are always available for use on the platform. Updates also ensure that your device and OS are protected against potential cybersecurity risks. We'll discuss the process of updating ChromeOS more in-depth in *Chapter 4, ChromeOS Security*.

Upgrade time

Although updates are a critical part of maintaining a healthy ChromeOS installation, there will come a time when they aren't enough. This occurs when Google releases a change to ChromeOS that is so drastic that a mere patch to the existing software won't suffice. This is when we need to perform an *upgrade*. An upgrade replaces the current version of ChromeOS with a newer version of the software, which incorporates all major enhancements and bug fixes. Some of the benefits of performing upgrades when they become available are as follows:

- Improved performance and stability
- Corrections to identified errors in the OS source code
- A new and enhanced feature experience

We'll discuss the process of performing ChromeOS upgrades in-depth in *Chapter 4, ChromeOS Security*.

Hardware flexibility with ChromeOS Flex

In previous versions of ChromeOS, the OS was bound to specific hardware. This meant in order to get access to ChromeOS, you had to purchase a Chromebook or Chromebox device. It also meant that as your device aged, it would eventually no longer be able to support the most up-to-date version of the OS. Although this isn't a problem unique to Chrome's OS, Google is one of the few vendors who have provided a solution other than purchasing upgraded hardware. That solution is ChromeOS Flex.

In 2022, Google released ChromeOS Flex as a way for people to continue to utilize aging computer hardware. What makes ChromeOS Flex so powerful is that it can be installed on any PC or Mac. Systems that were once bound for landfills and recycling plants are now able to be reborn as ChromeOS devices. Once ChromeOS Flex is installed onto a system, it provides the same functionality and user experience as a native ChromeOS device.

Summary

We have now come to the end of this chapter. We explored what makes Google's ChromeOS special when compared to other OSs on the market. We walked through the steps for setting up a brand-new ChromeOS device and customizing its functionality and aesthetics via the **Device** and **Personalization** settings menus. Finally, we introduced the concept of OS updates and upgrades, and explored how ChromeOS Flex can turn even outdated hardware into your new ChromeOS device.

In the next chapter, we will explore the many ways that ChromeOS allows you to connect to other devices and the internet.

2

Getting Connected

Although computing can be done on stand-alone devices with locally installed applications, our connected world has made this style of computing almost obsolete. In modern computing, connectivity is *key*. Everything from watching your favorite live stream to reading your favorite eBook on Packt's website requires connectivity. **ChromeOS** meets that need by offering multiple ways to connect your device to the world around you. Whether you're surfing the web, connecting wirelessly to your Bluetooth headphones, or sharing your screen with a TV on the other side of the room, the connectivity features available in ChromeOS will have you up and running with minimal effort.

In this chapter, we'll explore the different ways that you can use ChromeOS to connect to the world around you. You'll learn how to tap into wired, Wi-Fi, and mobile network connections to get online. You will also learn how you can ditch the wires when connecting to peripheral devices such as speakers and TVs. By the end of the chapter, you'll have a full understanding of how to get your ChromeOS device connected no matter the situation.

In this chapter, we'll be covering the following topics:

- Ethernet and Wi-Fi internet access
- Mobile data connectivity
- Bluetooth and connected devices
- Casting and syncing

Technical requirements

In order to follow along with the activities outlined in this chapter, you'll need access to the following services and technologies:

- Wired or wireless internet service
- Access to a mobile data card (built into your ChromeOS device or attached as a peripheral)
- Mobile network connectivity (such as internet provide by a cellular carrier)

- Bluetooth wireless device (such as a wireless mouse or keyboard)
- A ChromeOS-compatible TV or a **Google Chromecast** device

Ethernet and Wi-Fi internet access

As a cloud-first **operating system** (**OS**), ChromeOS relies heavily on access to the internet to provide users with the best computing experience possible. So, it's no wonder that getting connected to the internet is a top priority and is included in the initial setup process covered in *Chapter 1, ChromeOS Basics*. Here, we'll look at the two most common ways of getting online; wired internet connectivity via **Ethernet**, and wireless internet connectivity via **Wi-Fi**. We'll also explore how you can add a layer of security to your internet communication using **virtual private network** (**VPN**) technology.

Connecting with Ethernet

For many users, connecting ChromeOS to the internet with a wired Ethernet connection won't appear as a native option since the vast majority of Chromebook (laptop) hardware does not come equipped with a built-in Ethernet adapter. However, when using Chromebox (desktop) hardware or running ChromeOS Flex from a PC or Mac, you'll be able to natively take advantage of this feature. Additionally, there are several **USB-to-Ethernet adapters** in the market that can act as USB-attached Ethernet **network interface card** (**NIC**), giving even Chromebooks the ability to connect to wired networks.

Once you have the proper hardware in place, connecting your ChromeOS device to a wired internet connection is as simple as plugging an Ethernet cable into the ChromeOS device's Ethernet adapter port and ensuring the other end of the cable is connected to an internet source (such as a home router). Once connected, you will be able to confirm your internet connectivity by opening the **status tray**. There, as illustrated in *Figure 2.1*, you'll see a network connection type of **Ethernet** with a status of **Connected**.

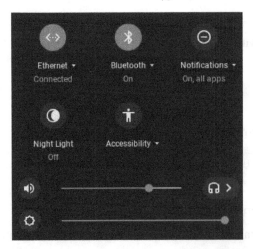

Figure 2.1 – The status tray showing the Ethernet network connection

Connecting with Wi-Fi

A more common way of connecting your ChromeOS devices to the internet is wireless via Wi-Fi. **Wireless NICs** are a standard component in all Chromebook and Chromebox hardware, making the use of Wi-Fi a convenient and reliable means of accessing the internet. To connect your ChromeOS device to the internet using Wi-Fi, follow these steps:

1. Navigate to the **Settings** menu. Under **Settings** → **Network**, select **Wi-Fi**.

2. If Wi-Fi is disabled, use the toggle button to enable it.

3. Once Wi-Fi has been enabled, you'll be able to click on the arrow to the left of the toggle button to view nearby Wi-Fi networks.

4. Select one of the Wi-Fi networks from the list and, if required, provide a password to connect to the wireless network.

The following figure provides an example of what you'll see when you access the **Network** → **Wi-Fi** settings menu.

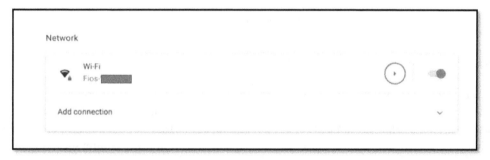

Figure 2.2 – The ChromeOS Network menu

As with the wired Ethernet network connection, you'll be able to verify network connectivity for your Wi-Fi connection using the status tray.

> **Pro-tip**
>
> In order to minimize unwanted access to their wireless networks, some organizations disable the broadcasting of the network's **service set identifier (SSID)**, which is essentially the name of the wireless network. By doing this, administrators prevent the network from appearing in the list of available networks during a computer's network search process. In order to connect to these hidden networks, you have to know their SSID and password in advance. Armed with this information, you can then use the **Add connection** option on the **Network** menu. The **Add connection** feature allows you to manually type in the SSID and password of the hidden wireless network. It also allows you to identify what type of **wireless security** the network uses. Finally, you're given the option of sharing the network you have identified with other devices.

Figure 2.3 displays the configuration window that you'll use to connect to hidden wireless networks.

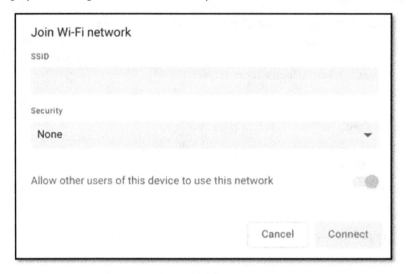

Figure 2.3 – The Network → Add connection window

Virtual private network (VPN)

Wired and wireless connectivity has many benefits, including allowing users the ability to connect to data sources around the world. However, that connectivity does present some drawbacks, namely the potential for our network communications to be captured and viewed by unauthorized third parties via the internet. Thankfully, we have tools that can help protect our network communications, one of the most important ones being VPNs.

VPN technology uses **encryption** to create a **secure tunnel** between a source and a destination device. This **encrypted tunnel** protects our data as it moves back and forth across the insecure internet. By implementing this security measure, we can be sure that even if our data is captured in transit, the contents of the communications will not be viewable to cyber criminals because they will not have the necessary *keys* to decrypt the data.

In order to implement a VPN solution, there are several third-party hardware and software options to choose from. Many of them require the installation of a small piece of software called an **agent** in order to take advantage of VPN's security features. ChromeOS is simplifying the implementation of VPNs by providing built-in agent software that uses widely popular **open source** VPN standards to ensure compatibility with many of the major VPN solutions on the market.

To connect your ChromeOS device to your VPN, perform the following steps:

1.   Navigate to the **Settings** menu. Under **Settings → Network**, expand the **Add connection** option.

2. Under **Add connection**, you will see the **Add built-in VPN** option. Click **plus sign** (+) to the right of the option to open the **Join VPN Network** screen.

3. In the **Join VPN Network** window, you are able to define the following settings:

    *   **Service Name** – The unique name that you'll use to identify the VPN connection (such as `Work VPN`).

    *   **Provider Type** – One of three VPN connection methods available natively in ChromeOS based on **Layer 2 Tunneling Protocol (L2TP)**, **Internet Protocol Security (IPSec)**, and **OpenVPN**:

        *   **L2TP/IPSec + pre-shared key**: A VPN connection method that uses a combination of the **L2TP** and **IPSec** protocols to provide end-to-end encryption for data as it travels between source and destination networks. The connection is secured using a secret key, which is shared in advance by the two parties that want to communicate.

        *   **L2TP/IPSec + user certificate**: A VPN connection method that uses a combination of the **L2TP** and **IPSec** protocols to provide end-to-end encryption for data as it travels between source and destination networks. This connection is secured using individual user security certificates that uniquely identify each user on a network. This VPN connection method is primarily used in business settings where user account information and credentials are centrally managed and maintained.

        *   **OpenVPN**: A open source connection protocol used by various VPN service providers. It can utilize pre-shared keys, user certificates, or even username/password combinations, locally defined on the **VPN host server**.

    *   **Server hostname** – The IP address or **hostname** of your VPN server.

    *   **Username** – The username of the VPN user.

    *   **Password** – The password of the VPN user.

    *   **Group name** – Used optionally by L2TP/IPsec VPN connections to define the IPSec identity field which some servers use to create *Tunnel Groups* or *User Realms*.

    *   **One-Time Pad (OTP)** – A single-use password generated by an **OTP card** or **VPN token**. This time-sensitive password option adds an additional layer of security for password-based OpenVPN connections.

    *   **Server Certificate Authority (CA)** certificate – A security mechanism that validates the identity of a server to a user. The setting is necessary when user certificates are employed to secure your VPN connection.

    *   **User Certificate** – A security mechanism used in user certificate-based VPN connections to uniquely identify individual network user accounts.

In *Figure 2.4*, you can see an example of the **Join VPN network** settings screen. Specifically, this figure displays the settings you'll configure when creating a VPN connection using the **L2TP/IPSec + pre-shared key** option.

Figure 2.4 – The Join VPN network configuration screen

In this section, we learned how ChromeOS connects to both wired and wireless networks. We also explored the built-in features the OS provides for connecting securely to remote networks via VPN.

Next, let's look at how we can connect to the internet while on the go using mobile data connections.

Mobile data connectivity

Although Ethernet and Wi-Fi internet connections are reliable means of connecting to the internet, they aren't always available. This is why ChromeOS gives you the option of wirelessly **tethering** your compatible mobile device to your ChromeOS hardware. This process allows you to share your **mobile data** connection with ChromeOS. In this section, we will explore the process of making this connection to mobile data sources.

Sharing the mobile data connection between your mobile device and ChromeOS device is an easy way to leverage vast cellular networks for internet connectivity. However, unless you have an unlimited data plan on your mobile device, you may incur significant costs because of the additional data usage. This is why Google doesn't recommend this connectivity method as a long-term internet solution.

A better approach for ChromeOS users on the go is to invest in a dedicated mobile plan for your ChromeOS device. Most major cellular carriers (AT&T, Verizon, Telefonica, etc.) allow you to purchase data plans specifically for devices such as Chromebooks. Depending on the make and model of your device, getting connected may involve installing a **SIM Card** into a special slot on your device, purchasing a **USB -to-SIM card dongle** to allow the SIM card to be attached externally, or simply purchasing a **USB modem** or **Wi-Fi hotspot** (*see Figure 2.5*).

Figure 2.5 – Several hardware solutions for connecting to mobile data

Regardless of what hardware solution you use, once you are able to connect to the internet using mobile data, the steps discussed in this chapter will have you surfing the internet in no time!

Built-in mobile data

If your ChromeOS device has the ability to use a mobile data plan directly from a cellular carrier via built-in hardware, then it further simplifies the already easy process of connecting to a **mobile broadband** network.

Use the following steps to set up your mobile data connection:

1. Click on **Status Tray** and select the **Settings** gear icon.

2. Under the **Network** section, select the **Mobile Data** option to turn it on:

    *   Note that you will need to consult your cellular carrier's documentation for the specifics of activating your mobile data plan.

    *   Also, note that ChromeOS will always choose to use a Wi-Fi connection over a mobile data connection whenever possible. Therefore, to force your system to use mobile data, you will need to turn off your Wi-Fi connection.

The following figure displays how the **Mobile data** option will appear on the screen in ChromeOS:

Figure 2.6 – The Network → Mobile data configuration screen

Mobile data connection using a dongle

A **dongle** is an adapter device that plugs into an open port on a computer in order to add additional functionality. In this case, the dongle plugs into an open USB port on your ChromeOS devices to allow you to externally connect a SIM card for mobile data access. In many cases, using the dongle simply requires plugging the device in. Once it powers up it operates just like an internal SIM card, providing you with an identical, simplified connection experience.

If, for some reason, your ChromeOS device does not automatically recognize the dongle, perform the following steps to manually configure it:

1. With your dongle plugged in, press *Ctrl + Alt*, then open the **crosh terminal** in your web browser.

    We'll discuss the crosh terminal in more depth in *Chapter 8, Working with Chrome Shell (Crosh)*.

2. Enter the following command: `set_cellular_ppp  -u  <userid>  -p <password>`:

    •   Replace `<userid>` and `<password>` with the login credentials assigned by your cellular provider

    •   If you receive a **No cellular service exists** error after running the command, reconnect the dongle, wait a few seconds, and then try again

3. After the dongle connects, go to **Settings** → **Wi-Fi** → **Mobile data** and enable your mobile data connection by clicking the **toggle switch**.

Mobile data connection using an Android device

In order to use an Android device such as a smartphone to connect to the internet via a mobile data connection, first confirm the following:

•   Your ChromeOS device supports **instant tethering**:

    •   You can view a list of devices that don't offer the instant tethering feature to ensure your ChromeOS device is not one of them: `https://www.chromium.org/chromium-os/chrome-os-systems-supporting-instant-tethering/`

•   Your device should have *ChromeOS version 71* or higher installed:

    •   You can verify this by navigating to the **Settings** menu and selecting **About ChromeOS**

•   If tethering to an Android device, your device should have *Android version 8.1* or higher installed

•   The Google account you're using on your Android device is the same account you use to log in to your ChromeOS device

Once you've met all of the requirements necessary to take advantage of this unique feature, perform the following steps to connect to your mobile data source:

1. Click on **Status Tray** and select the **Settings** gear icon.

2. Under the **Connected Devices** → **Android phone** section, click the **Set up** button.

3. When the **Connect to your phone** screen is displayed, as seen in *Figure 2.7*, ChromeOS will provide a list of all of the Android devices associated with the currently logged-in Google account. If one of the devices listed is one that you'd like to tether to, select the **Accept & Continue** button.

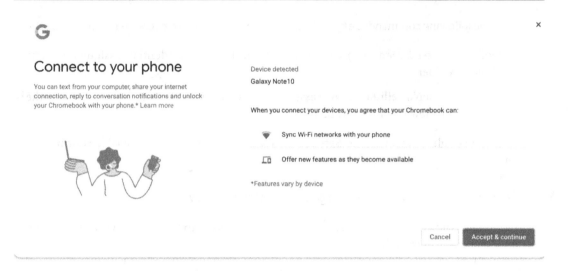

Figure 2.7 – The Connect to your phone screen

4. Next, you'll enter the password for your Google account and confirm the login on your Android device.

5. One the **All set!** screen appears, press the **Done** button to complete the tethering process.

6. Now, return to the **Settings → Connected devices** section and you'll see the newly tethered Android device listed as **Disabled**. Click on the toggle switch to the right of the option to enable the connection. *Figure 2.8* illustrates what you'll see on the screen once your connection is enabled.

    Note: Re-enter your password if prompted.

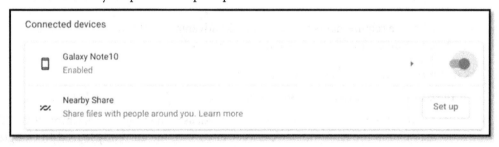

Figure 2.8 – Enabled connected device in ChromeOS

7. With your connected device enabled, you will now see the **Mobile data** option available under the **Settings → Network** section. Ensure your Wi-Fi is turned off and that your mobile data is turned on, then click on the arrow to the left of the **Mobile data toggle switch**, as seen in the following figure:

Figure 2.9 – The Mobile data feature enabled but not connected

8. After clicking on the arrow, you'll see all of your available mobile data connections. Click on the one you'd like to use to change its status to **Connected**, as shown in *Figure 2.10*. Now you're free to start surfing the web using mobile data.

Figure 2.10 – Mobile data with a Connected status

In this section, we covered all of the different ways ChromeOS can connect to mobile data networks. This provides the OS with the ability to connect to the internet even when an Ethernet or Wi-Fi connection isn't available.

In the next section, we'll take a look at how ChromeOS enables us to ditch the wires when connecting to peripheral devices.

Bluetooth and connected devices

When it comes to ChromeOS, there are a number of ways that devices can interact with the OS through external connections. These connections can be as simple as plugging a USB cable into an open USB port or connecting an **HDMI cable** from your Chromebook to your TV. However, with Bluetooth and Wi-Fi technology, you will find that many of your connections to peripheral devices can also be made wirelessly.

Connecting with Bluetooth

Bluetooth is a popular wireless technology used for connecting peripheral devices to one another over short distances (a few hundred feet). This has made the technology ideal for use with computer systems since they tend to stay within a close range of the add-on devices that they work with (printers, mice, speakers, etc.). Although most of the peripherals supported by Bluetooth can be connected via wired methods, the freedom of movement that Bluetooth provides makes the use of the technology a must.

To enable Bluetooth on your ChromeOS device, follow these steps:

1. Click on the **Status Tray** at the bottom right corner of your screen:

    If you see the **Bluetooth** option listed in the status menu, as shown in *Figure 2.11*, then your device supports Bluetooth:

Figure 2.11 – The Bluetooth connection button

2. Click on the **Bluetooth** icon to toggle the feature on.

3. Once Bluetooth is enabled, your ChromeOS device will begin to scan for nearby devices to pair with.

4. Once you detect a device with which you want to pair, select its name, and then follow the onscreen instructions to complete the pairing process.

5. To view the devices that have paired successfully with ChromeOS, go **Settings → Bluetooth** and click the arrow to expand the **Bluetooth** section. Once expanded, in the **Bluetooth** section, you'll see a list of all paired devices and their connection statuses. You'll also see a list of unpaired devices that were detected during the Bluetooth scan but are still waiting to be paired. An example of this can be seen in the following figure:

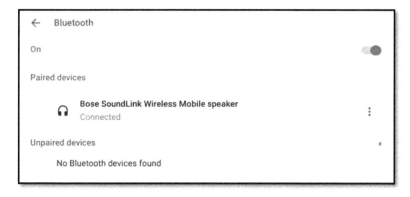

Figure 2.12 – The Paired devices and Unpaired devices lists

Connecting an external display

Although there are wireless methods for connecting displays to your ChromeOS device, wires still provide the easiest and most stable connection. Most ChromeOS devices also supports video connections, such as **HDMI** and **DisplayPort**, which allow you to have the superior video quality needed to view today's high definition, multimedia content. Connecting to an external display is covered in detail in *Chapter 1, ChromeOS Basics*.

Connecting with Chromecast or smart TV

ChromeOS provides an alternative to wired display connections. By using **Google's Chromecast** hardware or the hardware built into Android-compatible smart TVs, you can also wirelessly connect to the screens around you.

Google's Chromecast works by plugging into one of the free HDMI ports on your monitor or television for video connectivity and a USB port or power outlet for power. Once the device powers on, it is able to connect to your Ethernet (on select models) or Wi-Fi network. This connectivity enables it to communicate with computers and mobile devices that are connected to the same network. Your ChromeOS device, as well as many other devices, apps, and services, are then able to broadcast video and sound wirelessly to the display.

Android-compatible display devices (e.g. smart TVs) come with a feature called **Chromecast built-in**. As the name implies, these devices have Chromecast hardware and functionality built directly into the device. Instead of connecting a peripheral device to your TV, you simply use your remote to locate and enable the proper settings, and then you begin to **mirror** from your ChromeOS device to your TV or monitor.

In this section, we looked at the different ways we can create connections between our ChromeOS device and external displays. In the next section, you will learn how to send content to those displays and how to connect your ChromeOS device to other Android devices for a seamless user experience.

Casting and syncing

We've reviewed several methods for connecting your ChromeOS device to other technology, but to what end? Well, two of the best reasons for creating these connections are **casting** and **syncing**. In ChromeOS, casting is the process of **streaming** content from one device to another device using supported apps. Syncing is the process of creating a connection between your ChromeOS device, Android devices, and other devices that use the Google Chrome web browser. The sync process allows saved data, extensions, themes, search history, and other information stored by Chrome to be copied and synchronized across your devices.

Let's look at how to set up casting and syncing on your ChromeOS device.

Casting from Google Chrome

To cast from your ChromeOS device to a remote display, speaker, or streaming device (Chromecast, Roku, Amazon Firestick, etc.), you'll need to use the **Cast** feature built into the Google Chrome web browser:

1. On your ChromeOS device, open the Google Chrome web browser.

2. In the upper right-hand corner of the screen, select the **More** button (represented by three dots).

3. When the **More** menu opens, select the **Cast** option, as seen in the following figure:

Figure 2.13 – Google Chrome Cast option

4. Once you select **Cast**, you'll be presented with a list of compatible devices to cast to. You'll also be provided with a **Source** button, which you can use to either select casting a specific browser **tab** or your entire **desktop**. Select your device and source to begin casting. The following figure shows the casting menu:

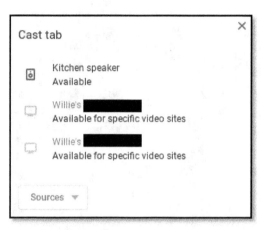

Figure 2.14 – The Cast tab menu

Casting from your desktop

In addition to casting from your Google Chrome web browser, you can also cast directly from ChromeOS's desktop to a Chromecast or Android-compatible smart TV. To use this feature, perform the following steps:

1. Click on the status tray in the lower right corner of the shelf.

    Use this as an opportunity to make sure that both **Wi-Fi** and **Bluetooth** are enabled.

2. On the status tray menu, select the **Cast** option.

    Note that if the **Cast** option isn't visible, it may be on the second page of the status tray. Simply click on the dots below the menu to move to the next page of options as shown in the following figure:

Figure 2.15 – Cast option on the status tray

3. Once the **Cast** option is selected, you'll see a list of all of the devices that are available to cast to. Select the desired device and begin sharing your screen.

Syncing with Android devices

Your Google account plays a central role in how you access resources in Google's technology ecosystem. Syncing ensures that regardless of what Google device you use to log in to your account, you can still have access to the settings, apps, and data you need.

To sync your ChromeOS device to an Android device, such as a smartphone, use the following steps:

1. Click on the status tray in the lower right corner of the shelf.

2. Select the **Settings** gear icon to open the **Settings** menu.

3. In the **Settings** menu, go to the **Accounts** section and select **Sync and Google services**, as shown in the following figure:

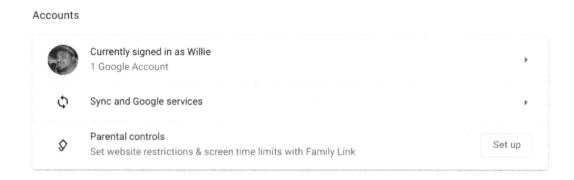

Figure 2.16 – The Accounts → Sync and Google services screen

4. Once selected, the **Sync and Google services** section will open, as seen in the following figure:

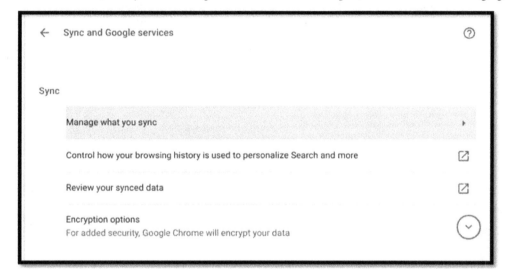

Figure 2.17 – The Sync options screen

The Sync menu options include the following:

- **Manage what you sync** – this option allows you to configure ChromeOS to **Sync everything** with your Android device or to create a **Customized Sync** that allows you to select specific data types to sync (bookmarks, apps, extensions, themes, and wallpapers).

- **Control how your browsing history is used to personalize Search and more** – this option redirects you to the **Google Account → Activity controls** webpage; https:// myactivity.google.com/activitycontrols. Here you can configure how account-related data is shared across all of your Google services. We'll take a closer look at these settings in *Chapter 3, Exploring Google Apps*, and *Chapter 4, ChromeOS Security*.

- **Review your synced data** – this option redirects you to the **Google Account → Chrome data in your account** web page: `https://chrome.google.com/sync`. Here, you are able to see a count of how many synced items you have in the Google account. We'll take a closer look at these settings in *Chapter 4, ChromeOS Security*.

- **Encryption options** – this option allows you to choose between two methods of data encryption in order to protect your Google account. We'll take a closer look at these settings in *Chapter 4, ChromeOS Security*.

Summary

In this chapter, you learned all about the different ways that you can connect to the world around you using ChromeOS. We explored the various methods of connecting to the internet, mastered connecting to peripheral devices using Bluetooth technology, and enhanced our user experience by casting to and syncing with other devices. Each of these distinctive capabilities helps ChromeOS to provide you with a user experience unmatched by any other OS on the market.

In our next chapter, we'll explore Google's app offerings.

3
Exploring Google Apps

Google offers a host of cloud-based applications that can be used natively by **ChromeOS**. Additionally, because of its integration with the Google Play Store, ChromeOS can also access thousands of additional apps. This chapter will explain the uses of the most common applications you'll find on ChromeOS, how to access them, and how to acquire additional applications.

You'll learn how to access and configure applications; what the most popular apps are that Google has to offer; how to use the Chrome browser to access even more app options; and how to get mobile apps from the Google Play Store onto a ChromeOS device.

In this chapter, we will cover the following main topics:

- Managing default apps and notifications
- Google Workspace applications
- Accessing Chrome browser-based apps
- Chrome Web Store
- Google Play Store

Technical requirements

In order to follow along with the activities outlined in this chapter, you'll need access to the following:

- A device with ChromeOS or ChromeOS Flex installed
- Wired or wireless internet

Managing default apps and notifications

As with any operating system, ChromeOS gives you access to a ton of software applications. However, one of the things that makes this cloud-first OS special is the many useful, built-in apps. In this section, we will look at some of the apps you can expect to work with on your ChromeOS device straight out of the box.

Additionally, with all of the innovative ways we can connect with ChromeOS's applications, managing **notifications** can get a little tricky. Therefore, to save you the frustration of dealing with too many notifications, we'll go over how to manage notifications the right way.

Once you're ready, let's unpack some default apps!

Default apps

Google does not skimp when it comes to providing users with the software tools they need for modern computing. Among ChromeOS's many apps are simple tools, such as calculators, and complex video conferencing apps, such as Google Meet. What makes an app different from traditional applications used by most desktop and laptop computers is its ability to run on resource-restricted hardware.

In terms of code and hardware resource requirements, apps tend to be much lighter and less demanding of resources (excluding **CPU** and **RAM**) than a traditional software application. It's for this reason that apps are the tools of choice for ChromeOS devices. Their light weight allows them to run powerfully on low-powered hardware. It also allows ChromeOS to come preloaded with a bunch of tools you'll need for work and play without forcing your system to grind to a halt.

As of the writing of this book, ChromeOS comes with the following default apps and core services:

- **Calculator**: A standard calculator program
- **Chrome**: Google's web browser program and the basis of many of ChromeOS's features
- **Chrome Canvas**: A basic drawing program
- **Docs**: A word processing app used to create complex text documents
- **Files**: Google file explorer app, used to help you navigate the file system
- **Gmail**: ChromeOS's default email application
- **Google Calendar**: A digital calendar application that integrates with Gmail and other apps
- **Google Chat**: An instant messaging and collaboration app
- **Google Drive**: A cloud-based storage app
- **Google Keep**: A note-taking app for creating multimedia memos and reminders
- **Google Maps**: An advanced GPS app that uses search features from Google's web browser
- **Google Meet**: A video conferencing application
- **Google TV**: A video streaming app that works along with a Chromecast or Android TV
- **Kindle Cloud Reader**: An app for reading Kindle e-books
- **Messages**: An SMS app for sending and receiving texts, photos, and voice messages
- **Photos**: A photo storage, organization, and sharing app

- **Play Books**: An app for accessing e-books, audiobooks, comic books, and manga
- **Play Games**: An app for accessing video and mobiles games from multiple genres
- **Play Store**: Google's app store, your portal for accessing thousands of Android apps
- **Sheets**: A spreadsheet app used to create tabular data
- **Slides**: A presentation app used to create and share slideshow presentations
- **Text**: A simple text editor app
- **Web Store**: An online store that provides extensions and web apps for the Chrome browser
- **YouTube**: – A video-sharing and social media app
- **YouTube Music**: A music streaming app

Although you will undoubtedly find many of these apps useful, there may be some that don't suit your needs. So, to help you tailor your app user experience, ChromeOS gives you the ability to easily manage all of your apps from the **Settings** menu.

To view and modify the full app list, perform the following steps:

1. Navigate to the **Settings** menu and select the **Apps** submenu, as seen in the following screenshot:

Apps

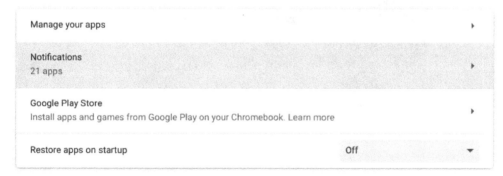

Figure 3.1 – ChromeOS Apps menu

2. From the **Apps** menu, select **Manage your apps**. This will allow you to view a full listing of the apps installed on your system, as seen in the following screenshot:

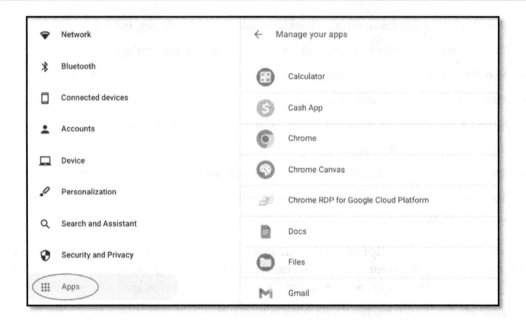

Figure 3.2 – The Manage your apps screen

3. On the **Manage your apps** screen, click on the app you want to customize to reveal its app-specific configurations. *Figure 3.3* illustrates some of the available app configuration options:

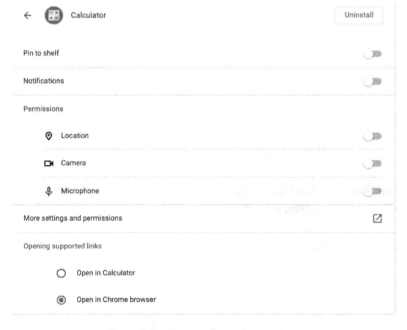

Figure 3.3 – App configuration options

Common default app configuration options include the following:

- **Uninstall**: This option allows you to remove unwanted apps from your system.

- **Pin to shelf**: This option provides you with a toggle that, when turned on, creates a shortcut to the app on ChromeOS Shelf.

- **Notifications**: This option can be used to enable or disable app notification messages on ChromeOS.

  Note that we will take a closer look at notifications in the next part of this chapter.

- **Permissions**: This section provides toggle switch options that allow or deny an app to access and use the following feature of your ChromeOS device:

  - **Locations**: This allows or denies an app the ability you use GPS to identify the location of a ChromeOS device.

  - **Camera**: This allows or denies an app the ability to access a device's camera hardware.

  - **Microphone**: This allows or denies an app the ability to access a device's internal or external microphone.

- **More settings and permissions**: This link redirects you to the advanced settings menu for the app. This menu contains an extensive list of options, giving the user an even more granular level of control over how an app functions.

- **Opening supported links**: This option allows you to set how any links related to the app will open. The options include the following:

  - **Open in app**: This forces the app configuration menu to open within the app itself

  - **Open in Chrome browser**: This forces the app configuration menu to open in the Chrome web browser

It's important to note that some configuration options will vary from app to app, so only the most common configuration options are covered in this book. For detailed instructions for specific app configuration, please consult the app vendor's documentation.

Now that you know how to properly manage the setting of your default apps, let's take a look at how notifications can help you stay up to date on each app's current activities.

Notifications

With so many apps included by default in ChromeOS, monitoring their activity can be a daunting task. What messages are they sending and receiving? What data are they processing? These are just a few of the questions that a ChromeOS systems administrator must answer when fine-tuning the operating system's functionality. Thankfully, Google has provided us with a group of settings that can be used to alert us of certain activities as they occur on our ChromeOS device. These are our **notification settings**.

In ChromeOS, notifications are used to alert users to things such as calendar invites, website activity, app communication, or browser extension activity. As mentioned earlier in this chapter, you can enable or disable notifications for specific apps by going directly to the app's configuration menu and using the toggle switch to enable or disable the feature. However, for centralized management of app notifications and access to advanced notification settings, you'll need to visit the **Notifications** menu.

To access the **Notification** menu, follow these steps:

1.   Navigate to the **Settings** menu and select the **Apps** submenu.

2.   From the **Apps** menu, select **Notifications** to access ChromeOS's centralized notification options, as seen in *Figure 3.4*.

Figure 3.4 – The Notifications menu

As you can see, the **Notifications** menu greatly simplifies the management of your app notification by putting each app's notification toggle switch on a single screen. Additionally, the **Notifications** menu provides users with the **Do not disturb** option, which can be used to quickly disable all app notifications with the click of a button.

Now that we have a handle on managing ChromeOS's default applications and their notifications, let's explore Google's popular cloud-based application suite, *Google Workspace*.

Google Workspace applications

Google Workspace (formerly known as G Suite) is a collection of cloud-based communication, collaboration, and business apps. Even though the software tools included in this app suite provide business-quality features, you get access to their core functionality at no cost to you. Simply set up a free Google account, and you're in! That's right, the same Google account you created to set up and use your ChromeOS devices also puts a treasure trove of software applications at your fingertips.

Google has invested a lot of time in helping to create and curate various software applications to meet just about any need a user might have. This is evidenced by the long list of apps that come with ChromeOS by default. However, not all apps are created equal. When it comes to everyday usefulness, **Google Workspace apps** stand above the rest. As Google's flagship suite of applications, they play an important role in making ChromeOS a viable solution for business, academic, and even personal usage.

Since ChromeOS and Google Workspace both use your Google account as your login credentials, accessing Google Workspace apps is a simple task.

Follow these steps to access your Google Workspace apps:

1. From the ChromeOS desktop, click the **Launcher** icon to open the **Launcher** menu.

2. Once the **Launcher** menu is open, expand it as shown in *Figure 3.5* in order to see all of the apps you have to choose from.

Figure 3.5 – Apps displayed by Launcher

Cloud-first doesn't only apply to the Chrome operating system but also to its apps. A major benefit of using Google Workspace apps is your ability to access them not only from your ChromeOS device but also from any device with an internet connection and web browser, and it doesn't even need to be Google Chrome. Just go to `Google.com`, log in to your Google account, and select the Google Apps launcher, as seen in the top right of *Figure 3.6*. How's that for convenience?

Figure 3.6 – The Google Apps launcher on Google.com

Now that you know how to access your Google Workspace apps, let's see what they can do.

Gmail

Gmail is one of Google's most recognizable web-based app offerings. So, it's no surprise that this popular email app is the cornerstone of Google Workspace. As one of the best web-based email services on the market, Gmail provides all of the features you would expect of a top-notch email solution. Additionally, through the integration of advanced features such as Google's web browser technology, the app has also become the most widely used web-based email application in the world. Gmail also acts as your key to unlocking a host of other free web apps since registering for Gmail webmail access simultaneously creates a Google account for your use.

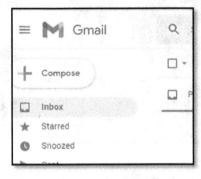

Figure 3.7 – Gmail

Drive

Drive is Google's cloud-based storage solution. It provides users with the ability to store, share, and collaborate on files and folders from any device with a web browser. If Gmail is the cornerstone of Google Workspace, then Drive is the backbone since it provides the storage service that all other Google Workspaces apps rely on. Here's an example of what you can expect to see when accessing Drive:

Figure 3.8 – Google Drive

Meet

Meet is the latest incarnation of Google's former video conference solution, *Hangouts*. Originally designed to be a paid service, Meet has evolved into one of the premiere, free video conferencing tools in use today. Its integration with other Workspace apps gives it additional advantages such as the ability to sync with your Google Calendar and to send invitations and updates seamlessly via Gmail.

Figure 3.9 – Google Meet

Calendar

Google Calendar is a powerful service that allows users to create, edit, and share events. The app provides all of the basic features of a digital calendar. However, through the inclusion of innovative tools and Google app integrations, Calendar is elevated to the status of a true time-management tool.

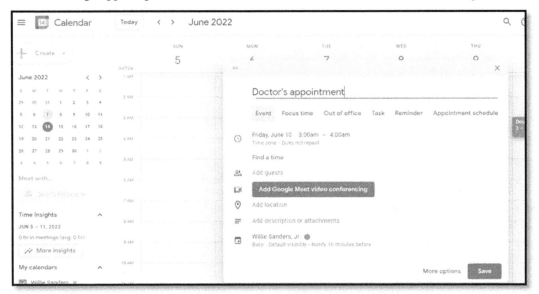

Figure 3.10 – Google Calendar

Chat

Similar to Google Meet, Google Chat was initially designed as a communication tool for business teams. However, it is now Google's primary tool for one-on-one and group-based real-time text communication. Chat also provides you with the ability to share documents from other Google apps, such as Docs, Sheets, and Slides via chat messages in order to facilitate conversation and collaboration.

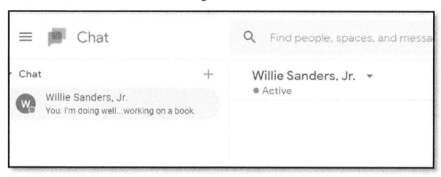

Figure 3.11 – Google Chat

Docs

Docs is Google's web-based word-processing application. It is used for creating rich, text-based documents but can also incorporate graphics, tables, charts, and other visual elements. In addition to providing cross-platform access for document creation, Docs allows users to collaborate in real time.

Figure 3.12 – Google Docs

Sheets

Sheets is Google's cloud-hosted spreadsheet application. This app is suitable for creating, editing, organizing, and analyzing tabular data. You can use it for creating budgets, tracking inventories, taking attendance, or any task that requires you to store data in a table. Sheets also takes your data management to the next level with the use of formulas to automate tasks.

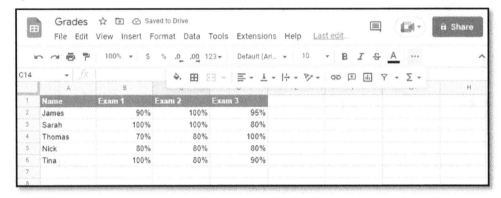

Figure 3.13 – Google Sheets

Slides

Slides is Google's web-based presentation application. The app enables users to create and format dynamic, visually appealing presentation documents, which can be shared across multiple platforms. The presentations can include pictures, video, and sound elements and also connect with other Google Workspace apps to incorporate their content.

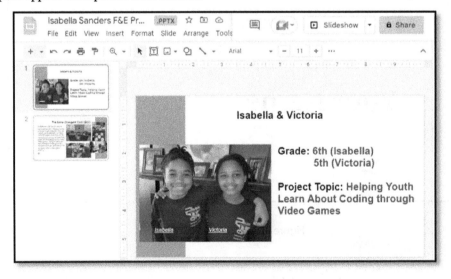

Figure 3.14 – Google Slides

Keep

Keep is Google's digital notebook and organizer application. Keep allows users to create different kinds of notes, including lists, text, images, and audio. Users can also connect and synchronize their notes from other apps so that they're accessible from a centralized location.

Figure 3.15 – Google Keep

Forms

Forms is Google's cloud-based web form tool. The app allows you to make online questionnaires, tests, registrations, or any other type of document you might need to collect information from your teams or the general public.

Figure 3.16 – Google Forms

Jamboard

Jamboard is Google's digital whiteboard app. This app allows users and teams to collaborate in real time on a shared digital canvas from their computer, mobile device, or even a specialized Jamboard device. You can sketch an idea, add a digital sticky note, and import a video clip from YouTube. Jamboard frees you not only to collaborate but to create.

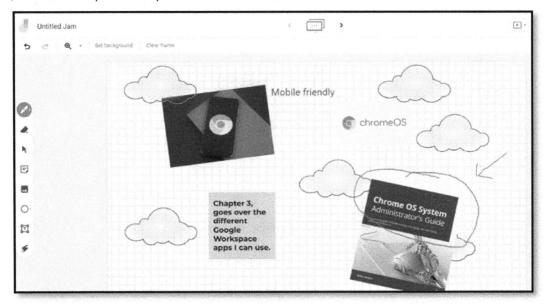

Figure 3.17 – Google Jamboard

Pro tip

You can take your Jamboard experience to the next level with a dedicated Jamboard device. The 55-inch, 4K, touch-sensitive smartboard creates instant engagement when used in meetings, classrooms, or any collaborative work setting. Additionally, since the physical device leverages Google's Jamboard web app, transitioning from the Jamboard device to your computer, tablet, or smartphone is seamless and instant. Note that the Jamboard device may not be available in all countries.

Figure 3.18 shows an example of the Google Jamboard device produced by the smartboard technology company BenQ®:

Figure 3.18 – Google Jamboard device by BenQ®

As you can see, Google Workspace offers ChromeOS users the ability to quickly access a huge selection of software tools to meet just about any need at no cost to you. But believe it or not, Google has even more apps to offer customers who decide to take advantage of their paid Google Workspace editions.

In the next section, we will take a look at what Google Workspace paid subscription plans have to offer.

> **Pro tip**
>
> Anyone who has dealt with technology in some capacity knows that it is an ever-changing moving target! So, as we discuss the structure of Google Workspace's premium subscriptions, be aware that they can change at any time without warning. That being said, the overall tiered approach to scaling access to services is a staple of the IT industry. So you can be sure that no matter what the details of each edition of Google Workspace are, there will undoubtedly be at least one option that suits your needs and your budget.

Freemium versus premium plans

Google has a long-running strategy when it comes to business: provide value upfront and monetize later. Because of this, many of Google's Workspace apps are **freemium** offerings. In the software world, freemium is a business model where a company provides basic or limited features to users at no cost and then charges a premium for additional resources or advanced features. All of the Google Workspace app offerings discussed in this chapter's previous section fall into the category of freemium apps.

But what happens when 15 GB of storage isn't enough space for all of the documents, pictures, and videos in Google Drive? Or when you want to hold a big virtual gathering, but your free version of Google Meet caps you at 100 participants with a 1-hour maximum meeting length? This is where Google Workspaces premium editions come in to save the day.

Pro tip

The most commonly used edition of Google Workspace is its personal edition. This is the free version you receive access to when you create a Google account or set up a free Gmail account. The apps and features included in this edition are phenomenal for personal use but may not provide the management tools you need for using Google Workspace in a business setting. But before you hang your head and prepare to break out your wallet, I have some good news. Google provides a few free editions designed for companies as well!

Premium tiers

Once you've decided to upgrade from Google Workspace's free edition to one of the paid subscriptions, you'll need to determine what services you really need and what level of resource capacity you'll need for each of them to have access to. Like all cloud services, Google Workspace follows a *pay for what you use* or utility-style pricing model. Therefore, the best way to ensure you're not overspending on your Google Workspace subscription is to carefully consider how you'll use the various apps it has to offer.

At the time of writing this book, Google provides several **Essential**, **Business**, **Education**, **Nonprofit**, and **Enterprise** editions to choose from. They include the following:

- **Business Starter**: This option provides access to all of the free edition Google apps but with increased cloud storage capacity, business-level Gmail access, and enhanced security and management controls. This option is one of the least expensive paid editions.

- **Business Standard**: This option has all of the features of Business Starter but with a whopping increase to 2 TB of cloud storage per user and an upgrade from 100 to 150 participants per Google Meet video conference. Expect to pay a little more per month per user for this edition… but not too much more.

- **Business Plus**: This edition adds to the Business Standard edition by further enhancing email security, providing 5 TB of cloud storage to users, and increasing user, endpoint, and user management capabilities. This is the highest level of the Business editions but still a cost-effective solution for a small to midsized organization.

- **Essentials Starter**: This edition excludes several of the apps you typically find in the free personal edition of Google Workspace. However, it makes up for these exclusions by adding in the centralized administrative controls and pooled cloud storage features those businesses with small teams need to thrive.

- **Enterprise**: This edition is fully customizable and tailored by Google to meet your organization's specific business needs. The cost for this edition varies widely since Google is building you a one-of-a-kind cloud solution. At this level, expect to spend a decent amount per month per user on services. However, you can rest assured that the services will be top-tier and best in class.

- **Education Fundamentals**: This is the first of several Workspace editions specifically designed to meet the needs of K-12 and higher education institutions. Education Fundamentals gives you access to core Google apps and services, plus others that have been deemed useful in a school environment.

- **Education Standard**: This edition adds to the Education Fundamentals edition by incorporating premium security and IT features. These features include a password vault app, enterprise endpoint management tools, and advanced email security.

- **Education Plus**: This edition builds on the features of the Education Standard edition by including, by default, an upgrade package called the **Teaching and Learning Upgrade**. This enhancement provides increased pooled storage, higher participant counts and more features in Google Meet, and access to a tool called **AppSheet**, which allows you to build no-code web apps.

- **Workspace of Nonprofits**: This is not so much an edition as it is a discount plan. Institutions that provide proof of nonprofit status are able to get access to Google Business Starter free with no cap on the number of users they can have. Additionally, they can receive a significant discount on other Google Workspace Business and Enterprise editions if they need to upgrade.

With your selection of the appropriate Google Workspace edition, you now have access to many of the apps you want and need. But wait, there's more! Next, let's take a closer look at some of the browser-based apps ChromeOS puts at our fingertips.

Accessing Chrome browser-based apps

In the early days of Google's cloud apps, when Google Workspace was still called G Suite, Google users relied primarily on the use of **Chrome apps**. A Chrome app differs from the **Google apps** we've been discussing thus far because of how it is deployed and accessed. Chrome apps, as the name implies, rely on the Chrome web browser in order to function. Chrome apps are essentially add-ons to Google Chrome, which allow additional features and functionality to be incorporated into the web browser.

A few good examples of Chrome apps are the **Kindle Cloud Reader** and **Chrome Remote Desktop**. Although these apps function like standalone applications because they are partially installed on your computer or mobile device, they, in fact, require the Chrome web browser to be installed on the system as well in order to run. *No Chrome browser, no Chrome app.* It's as simple as that. Google apps, on the other hand, are fully web-hosted services that exist independently of the Chrome browser. This gives them the freedom and flexibility to run on any web browser.

In 2016, Google announced that it would be ending its support of Chrome apps. Since then, it has removed Chrome apps from the Chrome Web Store (which we will discuss in the next section) and discontinued its support on the Windows, Linux, and Mac operating systems. Finally, in 2022, Google ended support for Chrome apps in ChromeOS as well. The apps will remain on your ChromeOS device but will no longer receive updates. We'll discuss why keeping apps that aren't updated is a bad idea in *Chapter 4, ChromeOS Security*.

To take one last look at Chrome apps on your ChromeOS device before they're gone for good, perform the following steps:

1. On your ChromeOS device, open your Chrome web browser and navigate to `https://chrome.google.com/webstore/category/extensions`.

2. In the left pane, select the **Apps** menu, as illustrated in *Figure 3.19*.

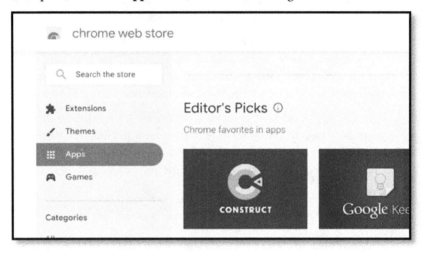

Figure 3.19 – Apps menu in the Chrome Web Store

3. On the right side of the screen, click the **View All** button to show all of the available Chrome apps.

Now that we've opened the Chrome Web Store, let's look around a bit.

The Chrome Web Store

The **Chrome Web Store** was once the one-stop shop for ChromeOS browser-based add-ons. Users could visit the web store to find apps, browser extensions, games, and themes. Each add-on helped to further personalize your user experience by providing them with additional software tools and useful features to further enhance the already feature-rich Chrome web browser.

As previously mentioned, Chrome apps are slowly being decommissioned by Google. So, even though you may find a few resources that are still available for download in the Chrome Web Store, most apps will indicate that they are no longer available for download. They will instead direct you to the app's website so that you can access the Google app version of the software, as seen in *Figure 3.20*.

Figure 3.20 – Chrome app no longer available

With the Chrome Web Store on its way to retirement, you may be thinking, *where am I going to get all of my cool apps now?* Well, if you are, there's no need to worry. Not only does ChromeOS have a replacement for the Chrome Web Store, but its replacement happens to be the largest app marketplace on the planet, the **Google Play Store**.

In the next section, we'll see how we can use ChromeOS to access the Google Play Store and its millions of app offerings.

Google Play Store

Google Play is Google's premier app store. It was originally designed to provide apps for use with Google's popular Android operating systems. However, with the transition away from browser-dependent Chrome apps and toward the **progressive web app** design of Google apps, the Google Play Store has become the go-to tool for adding more apps to the ChromeOS device. With over 2 million apps and more being added every day, there's no doubt that the Google Play Store has the app you need.

ChromeOS is able to take advantage of Google Play Store apps because both operating systems are derived from the Linux operating system. This common lineage has ensured that ChromeOS has the right code running behind the scenes to support Android apps. That being said, there are a few minor tweaks you'll need to perform before you can begin downloading your Google Play Store apps.

Follow these steps, and you'll be ready to play in no time!

1. From the ChromeOS shelf, click the Launcher and then select the **Settings** menu.

2. From the **Settings** menu, select **Apps**.

3. Under **Apps → Google Play Store**, click the **Turn on** button to enable **Google Play Store**, as shown in the following screenshot:

Apps

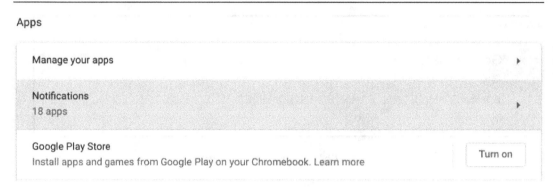

Figure 3.21 – Turning on Google Play Store

4. Once the button is pressed, the Google Play Store will prompt you to accept its Terms of Service. Press the **More** button to continue.

    Optionally, you can also opt to have your Google Drive automatically backed up and to use the locations feature by selecting the checkbox near the bottom of the window, as seen in *Figure 3.22*.

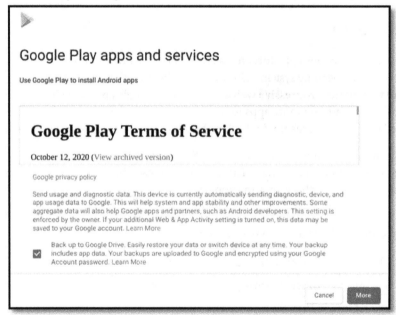

Figure 3.22 – Accepting the Google Play Store's Terms of Services

5. After selecting the **More** button, you'll see the last part of the app and services information. Press the **Accept** button to accept them and complete the activation process.

6. Once the activation completes, you'll be presented with the Google Play Store app, as illustrated in the following screenshot:

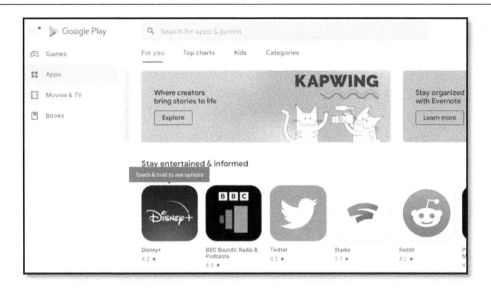

Figure 3.23 – The Google Play Store

And there you have it, Google Play on a ChromeOS device.

> **Pro tip**
>
> There are many ways to get apps onto your ChromeOS device, and we've touched on several. However, one more option that you might not be aware of is the Linux app store. Since ChromeOS is essentially a Linux operating system, it can handle running many of the same apps as a traditional Linux OS. We will discuss how you can use Linux on your ChromeOS device in-depth in *Chapter 7, The Linux Development Environment*, so be sure to keep reading!

Summary

In this chapter, we explored the default app and notification options available in ChromeOS. You learned about the default apps included in ChromeOS and how to access them. We discussed the various editions of Google Workspace and what unique benefits they each have to offer. Finally, you learned how Google has shifted its approach to providing apps, ultimately leveraging the Google Play Store as a major app repository for the OS.

Computer hardware is only as good as the software it supports. So, as you continue to explore ChromeOS in this book, use your newfound knowledge of how Google handles software to tap into the full potential of the Chrome operating system.

In the next chapter, you'll develop your understanding of how to secure ChromeOS from cyberattacks.

Part 2: Security and Troubleshooting

In this part, we will understand and implement the key networking and security features that allow systems running ChromeOS to communicate in a safe and secure manner.

This part comprises the following chapters:

- *Chapter 4, ChromeOS Security*
- *Chapter 5, Recovering from Disasters*
- *Chapter 6, Troubleshooting 101*

4

ChromeOS Security

In our digital world, being able to protect our data and the systems we use to access, store, and manage it is paramount. One of our first lines of defense in this ongoing battle to secure data is the operating system. So, it's no wonder that Google has invested heavily in designing the ChromeOS platform in order to make it not only highly functional but also highly secure. In fact, many argue that out-of-the-box ChromeOS is the most secure web browser on the market because it was designed with security in mind. This has led to many of its security features being standard components of the OS rather than added on after the fact.

However, even though ChromeOS is a streamlined and secure OS platform, it still has the potential to be compromised if the right protections aren't in place. In this chapter, we discuss the common ways you can further harden ChromeOS against cyberattacks.

In this chapter, we're going to cover the following main topics:

- User account management
- Screen locks and **two-factor authentication (2FA)**
- System updates, **sandboxing**, and **verified boot**
- Parental controls
- Chrome browser-based tools and settings

So, if you're ready, let's begin to take our ChromeOS defenses to the next level.

Technical requirements

In order to follow along with the activities outlined in this chapter, you'll need access to the following:

- A device with ChromeOS or ChromeOS Flex installed
- Wired or wireless internet

User account management

In cybersecurity, three main tenets are always on the minds of those tasked with defending systems and their data, and those who want to attack, steal, or destroy them. Those tenets are the following:

- **Confidentiality**—Ensuring that only authorized users have access to data or systems and no one else

- **Integrity**—Ensuring that only authorized changes can be made to systems and data, and only by authorized users

- **Availability**—Ensuring that data and systems are available to authorized users when they need them

These tenets are collectively referred to as the **CIA Triad**. Every security action that defenders or attackers take can be traced back to one or more of these fundamental concepts. **User account management** is a safeguard to confidentiality because it allows administrators to define who can access a ChromeOS system and what they are able to do once they have access.

Secondary user accounts

In *Chapter 1*, *ChromeOS Basics*, we worked through the process of creating our initial user account. However, ChromeOS isn't limited to having a single user per device. In order to create and manage additional ChromeOS user accounts, follow these steps:

1. Navigate to the **Settings** menu and select **Accounts**.

2. Under the **Accounts** menu, you should see the currently logged-in account listed. Click the account to expand its options menu, as illustrated in the following screenshot:

Figure 4.1 – Accounts menu

3. On expanding the **Accounts** menu, click the **Add Google Account** button, as seen in *Figure 4.2*. This will initiate the process of creating a new Google account or giving an existing account access to the ChromeOS device:

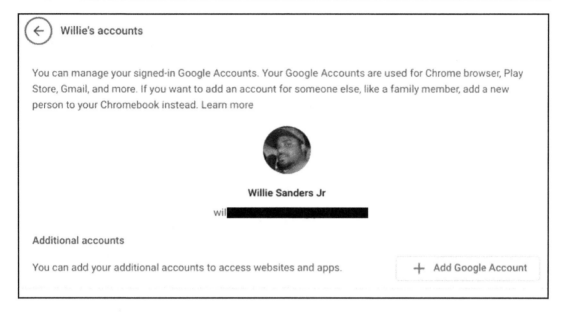

Figure 4.2 – Add Google Account option

> **Note**
>
> You may see a screen that details the purpose and benefit of adding another account to Google. Simply press the **OK** button to proceed to the account creation/login screen.

Once you've created a secondary user account on your system, all of the customization and personalization options discussed in *Chapter 1, ChromeOS Basics,* can be implemented.

Besides simply allowing you to create accounts, the **Accounts** menu also gives you control over how your Google account works across multiple devices and with other Google services. Although many of these settings prioritize functionality, they also play an important role in the overall security of the operating system, its key applications, and the data they store. To access these configurations, you'll navigate to **Settings → Accounts → Sync and Google Services**. The **Sync** menu allows you to synchronize all of your saved settings to your Google account. This means that if you log on to another device running ChromeOS, Android OS, or a Chrome web browser, all of your settings will be carried over. The **Other Google services** menu provides several advanced options to enhance the performance and security of other parts of ChromeOS and its core apps.

Now, let's take a closer look at each of these menus and their configurations.

Sync

The **Sync** menu provides four areas to configure, as pictured here:

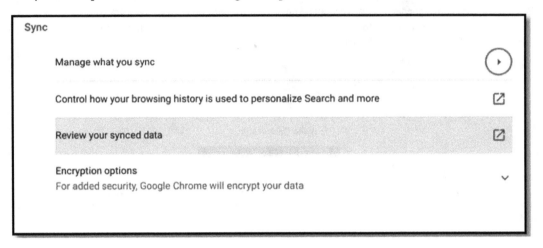

Figure 4.3 – Sync menu options

Under the **Sync** menu, you'll find the following configuration options:

- **Manage what you sync**—This option gives you the ability to either sync all of your ChromeOS device settings or create a customized list of items to sync. If customizing your sync list, you can choose from the following data sync options:

  - **Apps**
  - **Bookmarks**
  - **Extensions**
  - **History**
  - **Settings**
  - **Theme & Wallpaper**
  - **Reading list**
  - **Open tabs**

- **Passwords**

- **Address and more**

- **Payment methods, offers and address using Google Pay**

- **Wi-Fi networks**

- **Control how your browsing history is used to personalize Search and more**—This option redirects you to your Google account's **Web & App Activity** screen in the Chrome web browser. Here, you're able to take care of the following:

  - Turn on/turn off the **Save your activities on Google sites and apps** option

  - View/delete app activity

  - Include/exclude Chrome history and activity from sites, apps, and devices that use Google services

  - Include/exclude voice and audio activity

  - Customize auto-delete options

  - And more…

- **Review your synced data**—This option redirects you to your Google account's **Chrome data in your account** screen in the Chrome web browser. Here, you're able to see an actual count of how much data is being stored in your Google account by your apps, extensions, settings, autofill options, and other features. You are also provided with a **Clear Data** button that you can use to easily wipe this information from your Google account while allowing it to remain locally on your ChromeOS device.

- **Encryption options**—This option allows you to add extra security to your Google account by encrypting your data. You are provided with the following encryption options:

  - The **Encrypt synced passwords with your Google Account** option allows you to use Google's encryption methods to save and secure your password on its servers.

  - The **Encrypt synced data with your own sync passphrase** option also encrypts your data on Google's servers but it uses a passphrase that only you know. This means that no one besides you can read the data …not even Google.

Other Google services

The **Other Google services** menu also has a number of configuration options, as illustrated in the following screenshot:

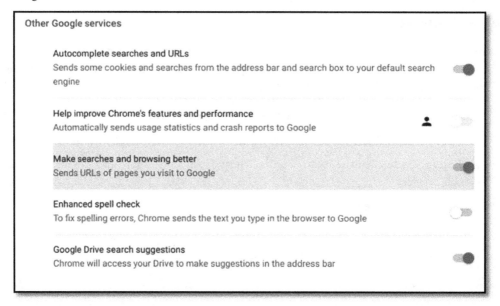

Figure 4.4 – Other Google services menu

Under this menu, you'll find the following configuration options:

- **Autocomplete searches and URLs**—If enabled, this option allows your Chrome web browser to perform quicker searches by creating more accurate search query predictions. These predictions are based on cookies and your search history.

- **Help improve Chrome's features and performance**—If enabled, this option allows Chrome to automatically send system performance data to Google. The data is then used to help Google prioritize which features and improvements to make to its products and services.

- **Make searches and browsing better**—If enabled, this option allows Google to anonymously collect usage statics on the websites you visit. This data is matched against a list of validated websites on your ChromeOS devices to ensure that your computer is surfing the web safely. If your Chrome web browser detects a site that hasn't been identified as a safe site, then the URL is flagged and sent to Google for further evaluation.

- **Enhanced spell check**—If enabled, this feature carries over the Google search engine's advanced spell-checking technology into the Chrome web browser.

- **Google Drive Search suggestions**—If enabled, this feature allows you to use the Chrome web browser's URL bar to obtain file suggestions for content stored in Google Drive.

Now that we have a better understanding of how the **Sync** and **Other Google services** account options can enhance account security and functionality, let's look at another—not so obvious—security feature of ChromeOS: **guest browsing**.

Guest browsing

In most modern operating systems, guest accounts are viewed in a negative light. For example, when administering a system running Microsoft's Windows OS, it's best practice to leave the guest account disabled. The reason for this precaution is that guest accounts don't require a username or a password to log in. For other operating systems, this is a major security red flag, but not for ChromeOS. In fact, guest browsing is a security enhancement!

When you utilize the guest browsing feature of ChromeOS, you are not required to enter your Google account login information into this system. Instead, you are able to access a temporary account to perform your computing tasks. This account has no access to any of the files and settings created by other Google accounts. You can surf the internet, use apps, and essentially use most of the ChromeOS device's features as you normally would. However, as soon as you end your computing session, the history of everything you did goes away. This ensures that your data can't be accessed without your permission when sharing a ChromeOS device with others.

You can easily access guest browsing as soon as you power on your ChromeOS device. Instead of logging in with your Google account, simply click the **Browse as Guest** button on the login screen, as seen in *Figure 4.5*:

Figure 4.5 – Browse as Guest login option

Even though guest browsing can be an extremely beneficial feature, you may have instances where it's not wanted or needed. To accommodate these ChromeOS requirements, administrators can perform the following steps to disable the **Browse as Guest** option:

1. Sign in to your ChromeOS device using the system owner's Google account.

    The system owner's account is the first account you created on your ChromeOS device.

2. Navigate to the **Settings** menu and select the **Security and Privacy** option.

3. Under the **Security and Privacy** menu, select **Manage other people**.

4. Finally, under the **Manage other people** menu, toggle off the **Enable Guest browsing** option, as illustrated in *Figure 4.6*:

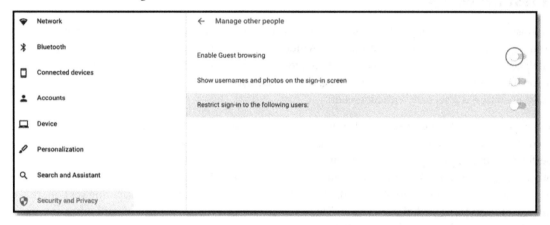

Figure 4.6 – Enabling/disabling guest browsing

Now that you have a better understanding of the benefits of guest user accounts, let's get an introduction to how businesses and organizations handle user account management for ChromeOS devices.

Advanced user management

Up to this point, we've been discussing user account management from a **decentralized** perspective, meaning that we've been focused on how to create and manage user accounts on individual ChromeOS devices. This approach is fine for your average home computer user or even business users in environments with limited numbers of users and devices. But what about ChromeOS devices deployed in enterprise environments? How do systems administrators maintain a cloud-first approach to administering devices while also maintaining the centralized control and management necessary in a corporate computing environment?

The answer is **Google Workspace for Enterprise**. If you recall, we introduced this subscription service in *Chapter 3, Exploring Google Apps*, within the context of the applications that the subscription type provides. However, **Google Workspace for Enterprise** subscriptions offer so much more…including advanced user account management tools you won't find in your basic ChromeOS installation. In *Chapter 10, Centralized Administration of OUs, Users, Groups, and Devices*, we'll take a deep dive into these advanced administration concepts.

So, now that you have your user accounts set up and somewhat protected, let's continue to add to confidentiality protections in our systems by learning to implement **screen locks** and **2FA**.

Screen locks and 2FA

When working in a modern computing environment, users will interact regularly with sensitive data and systems. Whether it's your personal bank account information or the schematics of your company's next big product, even a glimpse of this data by the prying eyes of a cyber-attacker can have catastrophic effects. This is why security-minded systems administrators follow a best practice called **Defense-in-Depth** (**DiD**, aka **layered defense**).

DiD/layered defense is a security concept that focuses on the strengthening of defenses through the implementation of multiple security measures instead of the hardening of a single security measure. Think of it like this: you want to protect your house from burglars, so you go out and buy the most expensive, high-tech door lock you can find. This lock does a fantastic job of protecting your front door. No one is getting in with it! But what about your windows? And the back door?

On the other hand, taking the DiD approach, you might opt for a mid-range door lock on the front door. However, you then combine that security measure with locks on your windows, a back door lock, security cameras, and a motion-sensitive flood light. Now, whose house do you think the burglars would rather attack? That's right—the one with the lowest number of defenses. This is true in physical security, and it's also true in system security.

In this section, we'll discuss how screen locks and 2FA can be used to provide another layer of defense for your ChromeOS device and users.

Screen locks

Screen locks play an important role in minimizing unauthorized access to your ChromeOS devices. In the same way that they stop someone from simply "swiping left" to get into your phone, screen locks require **re-authentication** using a PIN or password to regain access to your ChromeOS device once it's been locked. This comes in handy when you need to step away from your device for a bit but don't want to log completely out of the system.

By default, this feature is turned off on ChromeOS devices in order to allow you to quickly gain access to your system. However, it can easily be enabled by performing the following steps:

1. Navigate to the **Settings** → **Security and Privacy** menu.
2. Under the **Security and Privacy** menu, select the **Lock screen** option.
3. When prompted, confirm your Google account password in order to access the **Show lock screen when waking from sleep** and **Screen lock options** settings, as illustrated in *Figure 4.7*:

Figure 4.7 – Lock screen options

The lock screen menus options include:

- **Show lock screen when waking from sleep**—This option allows you to enable/disable the screen lock feature for inactive ChromeOS devices waking from sleep mode.

- **Password only**—This option requires you to re-enter your Google account password in order to unlock a locked system.

- **PIN or password**—This option allows you to create a PIN that you can use in place of your password to quickly unlock a locked system. You also still have the ability to use your password to unlock the system. Selecting this option will reveal the **Set up PIN** button, used to set your six-digit PIN.

4.   Once the screen lock feature is enabled with a password and/or PIN, use one of the following keyboard combinations to lock your system. The option you use will depend on your ChromeOS hardware:

- Press the **Search** key + *L*

- Press the **Launcher** button + *L*

Next, let's look at how 2FA can help to ensure you're the only one accessing your Google account.

2FA

Before you can really understand the importance of 2FA, you need to understand what a factor is in this context. As technologists have explored different strategies to protect the confidentiality of systems and data, they've come up with different ways of proving the identity of authorized users. These methods of proving your identity are referred to as factors, and there are several common factors that we use in computing, which include:

- Something you know (for example, password, PIN, security question, and so on)

- Something you have (for example, smart card, smartphone, USB key, and so on)

- Something you are (for example, fingerprint, facial recognition, and so on)

There are several other factors and implementations, but these are by far the most common. Also, these are the factors that Google has chosen to use in many of its 2FA solutions.

Up to this point, you've been utilizing authentication methods from the *something you know* category exclusively to prove your identity. The problem is this form of authentication is prone to compromise. People can guess weak passwords and even crack strong passwords using special software.

So, what do we do to protect ourselves? Well, we revisit our cybersecurity concept of DiD, of course! Only this time, the security layer we add is a second authentication method. By requiring more than one form of authentication, you exponentially increase the difficulty that a cybercriminal will have in compromising your account login.

As you implement a 2FA solution, though, it's important to note that in order for it to be effective, you must actually use two different factors. This means you have to use a combination of two of the three authentication methods previously stated (for example, *something you know*, such as a password, and *something you have*, such as a smartphone).

To implement 2FA on your ChromeOS device, perform the following steps:

1. Launch the Google Chrome web browser on your ChromeOS device.

2. Use the **Google Apps launcher** to access the **Account** app.

3. In the **Google Account** menu, select the **Security** menu.

4. Under **Security → Signing in to Google**, select the option to turn on two-step verification, as seen in the following screenshot:

Figure 4.8 – Accessing two-step verification

5. After clicking the option to enable two-step verification, you'll be guided through the process of registering a phone number to link to your Google account (if you haven't already done so) in order to enable your second authentication factor. Follow the onscreen directions to complete this process and finally enable 2FA.

With security measures such as screen locks and 2FA in place, your ChromeOS device is well on its way to becoming an impenetrable fortress! However, securing devices requires more than just protecting the login and authentication process. We also need to consider how the software running on the device will be protected from compromise.

In the next section, we will explore some of the methods ChromeOS uses to enhance software security.

System updates, sandboxing, and verified boot

ChromeOS's cloud-first design has major security benefits, including a reduced need to protect locally installed applications from compromise. However, a reduction of a security threat doesn't equal the elimination of that threat. Therefore, even ChromeOS needs tools to protect its software from being exploited by cyberattacks and malware.

In this section, we'll take a look at some of the tools ChromeOS supplies to help us prevent software compromise.

System updates

No matter how well a piece of software is designed, there will inevitably be some feature that was not included or some flaw that was overlooked. As software developers push the limits of their creativity and ingenuity to provide consumers with the latest and greatest software tools, things won't always work as intended. In the old days of software development, this would result in product delays and recalls. However, in modern computing, flawed code is typically fixed through a process called **patching**.

Just as a clothing patch is used to close up a hole in a fabric, software patches close up security holes in your application's code. Patches can also be used to enhance code by adding exciting new features and functionality. Although patching can be done using multiple methods, the most common way of deploying patches in ChromeOS is through **system updates**.

The System Update feature allows ChromeOS to reach out to Google's servers and download the latest ChromeOS updates allowed by your hardware. All other major operating systems provide similar functionality, but ChromeOS stands out from the crowd because of the ease of use and automation of the process.

Out of the box, ChromeOS has automatic updates enabled. This allows the operating system to automatically download the latest version of itself and install it quietly in the background. The process doesn't even force you to reboot! It just waits until the next time you choose to restart your system.

Additionally, updates can be done on demand. To perform a manual system update on your ChromeOS devices, follow these steps:

1. Navigate to the **Settings** menu and select the **About ChromeOS** option.

2. On the **About ChromeOS** screen, you will see a **Check for updates** button, as illustrated in *Figure 4.9*. Press it to force ChromeOS to check for updates to the operating system:

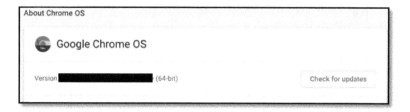

Figure 4.9 – Checking for ChromeOS updates

Note that if your Chromebook or Chromebox hardware has reached its **end of life** (EoL), you will not see the option to check for updates. Also, note that if you select the **Additional details** option after clicking **Check for updates**, you can see the date when your system will be eligible for its final update before going EoL.

3. After the **Check for updates** button has been pressed, ChromeOS will download and install any available updates.

Unfortunately, **system updates** only update the ChromeOS and Chrome browser software. Additional apps that you have downloaded from the Google Play Store will need to be updated separately. The good news is that updates for these apps can also be automated with a few simple steps. To enable automatic updates for Google Play apps, follow these steps:

1. On the ChromeOS Shelf, click the **Launcher** button.

2. Click the up arrow to view all of the available apps and then select **Google Play**.

3. Once you've launched the Google Play app, click on your account icon in the upper-left corner of the screen. This will cause the Play Store management screen to open, as seen in *Figure 4.10*. Select the **Settings** option:

Figure 4.10 – Play Store management options

4. From **Settings**, select **Network Preferences → Auto-update apps**, as illustrated in the following screenshot:

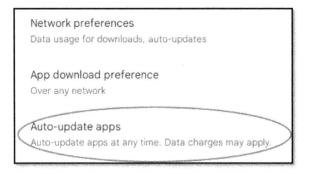

Figure 4.11 – Google Play Auto-update apps setting

5. The **Auto-update apps** setting provides you with the following options to choose from:

    *   **Over any network**—This option will allow automatic updates to occur on your ChromeOS device using either Wi-Fi or mobile data. Note that if your device is using mobile data, you may incur additional data usage charges.

    *   **Don't auto-update apps**—This option disables auto-updates for Google Play apps.

> **Pro tip**
>
> To continue using the **Auto-Update apps** feature without incurring extra mobile data charges, navigate to **Settings à Network Preferences à App download preference** in the Play Store. Here, you'll have the options to configure updates to only download when you're connected to Wi-Fi or to be prompted for your preference before every download attempt. Google's typical release schedule for updates is every 4 weeks.

Updates help to ensure you have the latest and most secure version of your operating system and its associated apps. However, in the event that an app does present a security risk, ChromeOS has another key **safeguard** in place that has the power to block rogue applications from affecting your entire computer. In the next section, we will explore the concept of **sandboxing** and learn how it's implemented in ChromeOS.

Sandboxing

In information technology, **sandboxing** is the process of isolating a piece of software in order to limit which hardware and software resources it is able to interact with. By doing this, you provide the application with only what it needs to do its job and nothing more. Additionally, you create an environment where the activities of a single software application are not able to negatively affect the system overall.

In ChromeOS, every web page that you open in the Chrome browser and every app that you launch creates its own isolated sandbox. This means that if an infected app makes its way onto your device or if you visit a website that has been hacked, you don't have to worry as much. The issue, by default, has already been contained and eliminating it is as simple as closing the infected app or the browser tab of the hacked website.

Next, let's look at how ChromeOS ensures a safe and secure computing environment from the very moment you press the power button using a feature called **verified boot**.

Verified boot

In order to protect the integrity of your system and its configurations, ChromeOS performs a process called **verified boot** during every system startup. During the verification process, the digital signatures of various core software components are checked to ensure that Google has approved them. The components include **firmware**, the **OS kernel**, the **initial RAM disk (initrd)**, the **master boot record (MBR)**, and so on.

Your ChromeOS device will only boot up if each of the checked items is successfully verified. If any of them fail this integrity check, if malware is detected, or if any other suspicious activity takes place, the boot process is halted. The system is then loaded into **Recovery** mode, which will allow an administrator to install healthy copies of ChromeOS's core system files.

> **Pro tip**
>
> For systems administrators, having tools ready for recovering the systems you support is a must. In order to recover a corrupted Chrome operating system, the tool you'll need is a **recovery storage disk (RSD)** for the make and model of the ChromeOS devices you need to repair. For detailed information on how to create an RSD for your ChromeOS device, please visit `https://support.google.com/chromebook/answer/1080595?hl=en #zippy=%2Cstep-download-a-new-copy-of-the-os%2Coptional-reuse-your-usb-flash-drive`.

Now that your hardware and OS are safe and secure, we can focus on keeping on little ones safe and secure online. For that, there's no better tool than **parental controls**. The next section will look at how to implement them to safeguard our children from inappropriate content.

Parental controls

As responsible administrators, we should always do our best to provide a safe computing environment for our users. From a technical perspective, this may involve implementing many of the security measures previously discussed in this chapter. However, protection from inappropriate content can be just as important, especially when we administer systems for younger users.

To assist us with this, ChromeOS provides a built-in **parental control** tool called **Family Link**. For Android OS users, this tool should seem very familiar because it's the same app used to provide oversight on your young person's Android-compatible smartphone or tablet. When **Family Link** is connected to a child's Google account, the **Family Link** manager (typically a parent) can control which apps can be downloaded, which websites can be viewed, how long a device is allowed to be used, and much more.

To access the parental controls in ChromeOS, perform the following steps:

1. Navigate to the **Settings** menu and locate **Accounts à Parental Controls**.

    In the **Parental controls** section, click the **Set up** button, as seen in *Figure 4.12*:

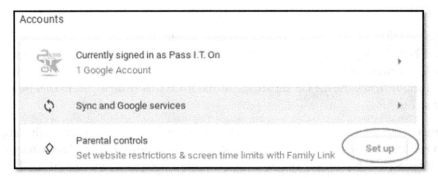

Figure 4.12 – Parental controls Set up button

2.   Once the **Set up parental controls with Google's Family Link** page opens, click the **Get started** button at the bottom of the screen.

3.   The **How to set up parental controls** screen provides an overview of the steps you will need to take to begin managing your child's device. After reading this page, click the **Next** button to continue.

4.   On the **Is this the child you want to supervise** screen, choose the account you wish to apply the parental controls to and press the **Yes** button.

Please note that in order to perform this step, your child must have an account on the devices you're currently working on and you must be logged in with that account. If the child has an existing account but you're not logged in with it, simply log out of the system and log back in with the child's account. If the child doesn't have an account, you'll need to create one before proceeding.

5.   On the **How parental controls work on your device** screen, you receive a summary of devices that can and can't be supervised based on your Google account login as well as the device's make and model. Click the **Next** button to continue.

6.   On the **Parent account** screen, enter the email address of the Google account that will act as **Family Link** manager, as illustrated in *Figure 4.13*, and then press the **Next** button:

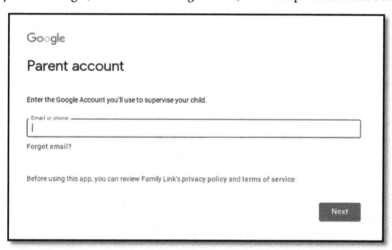

Figure 4.13 – Parent account login screen

7.   After the parent account email is recognized, you'll be prompted for your password. Enter it and then press the **Next** button.

Note that you may have to perform **2FA** if you've enabled that security feature.

8.   The **About supervision** screen provides a detailed overview of what parents can and can't see and do once **Family Link** is enabled. According to the onscreen instructions, parents are highly

encouraged to review this page with their young person prior to finalizing the parental control activation. Once the details have been reviewed, enter the password for the child's account to agree to supervision and then press the **Agree** button to continue.

9. ChromeOS will then complete the process by confirming that your accounts are linked and then providing you with a review screen that lists the apps that are currently installed on the device along with their maturity rating, as seen in *Figure 4.14*:

Figure 4.14 – App review screen

In addition to listing the apps on your device, the review screen also provides the ability to allow or block apps from being accessed by the child's account. Simply click the toggle switch associated with an app to allow or block it.

10. On the **Manage filters & settings** screen, you can adjust **Family Link**'s default filter to allow or deny certain types of content for different apps and tools in ChromeOS and the Chrome browser. You can customize the following items and then click the **Next** button to continue:

- **Filters on Google Chrome**—Allows you to block access to mature websites

- **Filters on Google Search**—Allows you to block explicit search results

- **Google Play app and games**—Allows you to change the maturity level of apps that are allowed to be downloaded from the Play Store (for example, **Everyone**, **Everyone 10+**, **Teen**, and so on)

- **Google Play movies**—Allows you to change the allowed rating level for movies viewed with Google Play movies (for example, **G**, **PG-13**, and so on)

- **Google Play TV**—Allows you to change the allowed rating level for TV shows watched with Google Play TV (for example, **TV-Y, TV-Y7**, and so on)

- **Google Play books**—Allows you to block sexually explicit books from Google Play books

- **Google Play purchase and download approvals**—Allows you to adjust which purchases require parental consent in Google Play

- **Controls for signing in**—Allows you to require permission when a child attempt to log in to the Chrome browser on a device that can't be supervised directly by **Family Link** (for example, Apple iPad and Apple iPhone)

11. On the **Supervise from your own device** screen, you are informed that you can install the **Family Link** app on your ChromeOS or Android OS devices in order to manage your child's parental controls. Click the **Next** button to continue.

12. On the **Automatically install Family Link on your devices** screen, you are presented with a list of all devices that you've logged into with your Google account. Here, you have the option of selecting the checkboxes of each device that you'd like **Family Link** installed on and then allowing Google to perform the installation process for you.

    Alternatively, you can skip this step and manually install **Family Link** on your other devices.

After finalizing these steps, you can rest easy knowing that the content being accessed on your ChromeOS device is kid friendly.

Now, let's shift our focus to the Chrome browser to see how we can make it more secure for users of all ages.

Chrome browser-based settings

As you know, the Chrome browser is a critical component of ChromeOS. It provides the core functionality around which the entire operating system revolves. Therefore, it goes without saying that it must also be secured against cyberattacks. Many of the security features previously mentioned in the chapter have an effect on the security of the Chrome browser. However, there are several additional tools and features incorporated into the browser itself to harden it against attacks. These tools focus primarily on protecting the **confidentiality** and **privacy** of the browser's users and their data.

To access the Chrome browser's security settings, follow these steps:

1. Launch the Chrome browser.

2. In the upper-right corner of the browser window, click on the three dots to view the **Customize and Control** menu.

3. Select the **Settings** option from the menu.

4. On the Chrome browser **Settings** page, select the **Privacy and Security** option.

5.   Under **Privacy and security**, as seen in *Figure 4.15*, you'll find various tools:

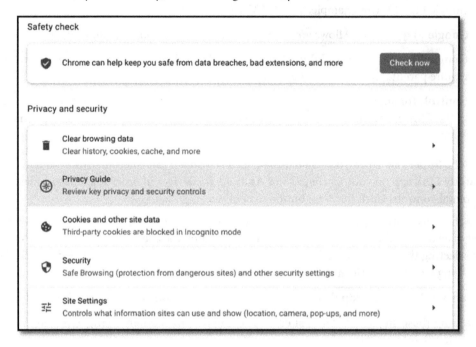

Figure 4.15 – Chrome browser privacy and security settings

These tools include the following:

- **Safety check**—This tool runs an automated vulnerability check of your Chrome browser. It then reports if there are missing updates, compromised or weak passwords, harmful extensions, or unsafe browsing practices in use.

- **Clear browsing data**—This utility provides several basic and advanced options for clearing cached browser data including **history**, **cookies**, **cached images**, and **files**. The tool also allows you to specify how far back in time you like to clear the browser, starting at **1 hour** and growing to **All time**.

- **Privacy Guide**—This tool provides a wizard that walks you through a review of several key privacy settings.

- **Cookies and other site data**—This feature allows you to define how website cookies should be handled (for example, allowed or blocked). It also lets you define cookie clearing and tracking rules.

- **Security**—This feature allows you to choose between three preset collections of browser security settings. They range from no protection to enhanced protection. The menu also provides some advanced settings for managing security certificates, keys, and DNS settings.

- **Site Settings**—This feature allows you to configure and control what information websites can show and use. This is where you go to enable or disable a site's ability to access your location data, camera, and microphone, among other things.

With the addition of these browser-based security settings, ChromeOS is able to present a well-rounded, layered defense approach. No wonder it's considered by many to be the safest OS on the market.

Summary

In this chapter, we took a deep dive into the many built-in security features that ChromeOS has to offer. After gaining a better understanding of key cybersecurity concepts such as the CIA Triad, we determined how ChromeOS applies the concept to user access control. You learned how to implement screen locks and 2FA to help harden the authentication process on ChromeOS devices.

You examined the system update process, sandboxing, and verified boot as mechanisms for protecting hardware and core software components. You set up parental controls in order to prevent unwanted content from getting into the hands of kids. Finally, you learned to identify and configure key Chrome browser settings in order to protect your data and privacy when online.

In the next chapter, you'll discover what you can do to recover your data following a digital disaster.

5

Recovering from Disasters

As you've learned from the previous chapters of this book, both hardware and software play important roles in making ChromeOS a cutting-edge operating system. But at the end of the day, it's not the computer system or its software that you or the users you support care about...it's the data. In our digital world, data is king!

Google's understanding of the true value of data has led them to design ChromeOS with several built-in data protection features. Users can leverage these features to ensure the security of not only their data but their apps as well. These protections can help to provide peace of mind for our most important and sensitive data assets. Additionally, if your data becomes totally unavailable due to a physical or technical disaster scenario, ChromeOS provides **Disaster Recovery**.

In this chapter, you'll learn about the following:

- ChromeOS's default data encryption technologies
- How to implement a data backup strategy using Google Drive or external storage
- How to locate and utilize OS and app recovery tools

So, if you're ready, let's start building our data defense and recovery skills!

Technical requirements

In order to follow along with the activities outlined in this chapter, you'll need access to the following:

- A device with ChromeOS or ChromeOS Flex installed
- Wired or wireless internet
- An external storage device with 8 GB or more of storage space (for example, a USB drive, external hard drive, or SD card)

Hardware and cloud-based data encryption

In modern computing, there are a number of methods to protect the precious data that flows through our computer systems. However, very few can match the effectiveness of **encryption**. Although other operating systems leverage encryption as a protective **countermeasure** to certain cyber attacks, ChromeOS incorporates this powerful technology in a unique, streamed-lined, and simplistic way.

In this section, we will learn about the basics of encryption and see how ChromeOS uses it to protect the confidentiality of your data.

What is encryption?

Encryption can be a complex topic. In fact, there are entire books written on the subject. A brief search of Packt's website is proof of that! Thankfully, you don't need to be an expert in encryption to take advantage of its benefits when using ChromeOS. However, as an administrator, you should still understand the basics…so let's dive in.

Simply put, encryption is a way of scrambling data so that only authorized individuals can understand it. It may sound high-tech but this science isn't anything new. In fact, the earliest evidence of encryption dates as far back as 4,000 BC in ancient Egypt. Even Julius Caesar used encryption during his reign over the Roman Empire to obscure battlefield communications, ensuring that if an enemy captured any of his correspondence, they wouldn't be able to read it. *Figure 5.1* shows a Roman cipher disk, one of the cryptographic tools used by ancient code makers.

Figure 5.1 – Cipher disk

Today, encryption works in much the same way as it did thousands of years ago. The process begins with your data, referred to as **Plaintext**, being sent through a complex math problem or process called a **Cryptographic Algorithm** (also known as a **Cipher**). The cryptographic algorithms themselves are publicly available for anyone to use. What makes them secure is the process of combining them with a **Cryptographic Key**. This key is essentially a secret code that only you and the people you want to communicate with know. When combined with a key, a cryptographic algorithm can alter plaintext in a way that appears random but secretly contains a complex pattern that can be reversed, if you know the key to reverse it.

The data created as a result of the encryption process is called **Ciphertext**. *Figure 5.2* illustrates the previously described encryption process.

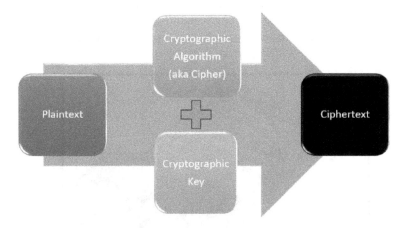

Figure 5.2 – Data encryption process

On the receiving end of the communication, this process will be performed in reverse in order to **decrypt** your data. Decryption involves taking your ciphertext and putting it back through the same cipher/key combination. The result is the restoration of the original plaintext message.

Now that you understand the encryption process a little better, let's see how Google has implemented it in the Chrome operating system to safeguard your data.

ChromeOS hardware-based encryption

In computer technology, you'll find that data can exist in three states:

- **Data in motion**
- **Data in use**
- **Data at rest**

Data in motion describes data moving between the subsystems of your computer device or across the network. Data in use describes data that is being created, modified, read, or processed by an application, a database, RAM, or a CPU. Data at rest describes data that is stored locally or remotely on hard drives, USB drives, cloud storage, archival tape, and so on. Encryption can be implemented to protect data in each one of these states but the type of encryption will vary.

Hardware-based encryption uses specially designed components to protect data at rest and data in use on a computer system. Many computer and mobile device manufacturers optionally leverage hardware-based encryption but all ChromeOS-compatible hardware comes with the ability to provide hardware-based encryption activated by default. It's a hardware requirement! This means, out of the box, your ChromeOS devices protect your data via encryption.

The primary component responsible for managing this security control is the **Trusted Platform Module** (**TPM**) chip. The TPM chip, like the one seen in *Figure 5.3*, is a specialized cryptographic processor, designed to handle all of the cryptographic (encryption and decryption) operations for a computer system.

Figure 5.3 – TPM chip

Additionally, in ChromeOS this hardware component provides some other important security services, including the following:

- Preventing software and firmware version rollbacks
- Protecting user data encryption keys
- Providing evidence of device tampering

The TMP chips in ChromeOS devices utilize the **128-bit AES encryption** standard. This is the same encryption standard used by the United States government to protect their most sensitive data. In fact, the encryption keys produced by 128-bit AES are so complex that researchers have determined it would take a supercomputer 1 billion billion (yes, that is a real number) years to crack the encryption keys it produces using software-based password-guessing techniques. Therefore, with hardware-based encryption, you can be assured that if your ChromeOS device is lost or stolen, the data locally stored on your device won't be accessible.

Next, we'll look at the role software-based encryption plays in ChromeOS's data security strategy.

ChromeOS software-based encryption

So now you're probably thinking, "With all of the data security provided by hardware-based encryption, I don't need anything else." Well, this is only partially correct. There is no doubt that hardware-based security provides powerful protection. However, hardware is no good without the proper software to help it carry out its purpose. Additionally, if you recall from our discussion in *Chapter 3*, *Exploring Google Apps*, the vast majority of your data is not stored on your local hardware in ChromeOS devices; it's stored in the cloud. This means that the software we use to connect to and interact with our cloud resources needs to provide encryption mechanisms capable of protecting our data in its various states as well.

In ChromeOS, software-based security begins at the filesystem level. The operating system takes advantage of its Linux foundation by leveraging the **ext4** filesystem's **fscrypt** encryption functionality. One of the main benefits of the fscrypt command library is that it allows the filesystem to support the automatic encryption of individual files and folders. Each file and folder also utilizes its own unique cryptographic key. The outcome of this feature is the **Sandboxing** functionality discussed in *Chapter 4*, *ChromeOS Security*.

In a nutshell, the ext4 filesystem enables each ChromeOS user directory to act as a vault for a user's data and encryption keys. These vault/key combinations are associated with each user's Google account. This means that only someone who knows the username and password of the account can access the data within. This isolation of user data is one of the major security benefits that ChromeOS boasts when compared to other popular operating systems.

Google Drive also plays an important role in ChromeOS's software-based security strategy. Due to its cloud-first approach, the majority of apps used on ChromeOS systems default to using Google Drive as their primary data storage location. The benefit here is that Google Drive provides some major protective measures as part of its design. For starters, Google Drive provides **256-bit AES** encryption to protect data at rest. It also protects data in motion by using a technology called **Transport Layer Security** (**TLS**). The TLS protocol is used to protect data from being intercepted and read by unauthorized users as it leaves your computer system and travels across the internet.

Lastly, the **Sync** menu, which is detailed in *Chapter 4*, *ChromeOS Security*, also helps to protect the Chrome browser's data that is synced with your ChromeOS device by adding encryption as an additional layer of defense. The encryption options illustrated in the following screenshot allow you to add additional security to your synced passwords stored on Google's cloud servers by encrypting with either your username and password or a unique sync passphrase.

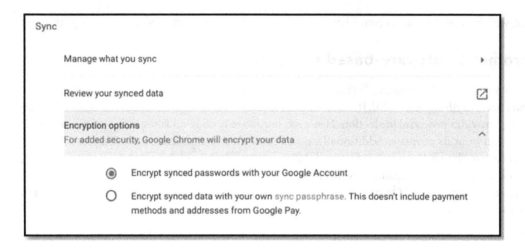

Figure 5.4 – Sync menu encryption options

In the next section, we'll discuss how backups are used in ChromeOS.

Data backup strategies for ChromeOS

Even with multiple safeguards in place to protect your data, bad things can still happen. There will be times when your data and the systems that house it are lost, corrupted, and become inaccessible. It's during these times that an administrator has to rely upon one of the most tried and true data disaster recovery tools – the backup.

In this section, we'll discuss the strategies and tools that you can take advantage of to ensure that you can get back lost data following a disaster scenario.

Backup basics

Performing system and data backups are very common tasks for both users and administrators. In fact, restoring data from a backup is the last-resort solution for resolving a myriad of data availability issues. But what is a backup? Simply put, a backup is a copy of data that is stored in an alternative location so it can be recovered if the original becomes unavailable. Over time, many different methods for performing backups have been developed. Additionally, the "alternative locations" that users and companies have identified for storing data have also evolved.

There are different backup strategies used to address the needs of different computer systems and computing environments. Each strategy requires an administrator to answer questions such as the following:

- How often do I want to back up my data? Every hour, once a week, once a month, or at another frequency?

- What data should I back up? All user data, mission-critical files, or folders only?

- Where do I want to store my data backups? In my office, at a remote facility, or in the cloud?

- What media do I want to store my backups on? An external hard drive, a **Network-Attached Storage (NAS)** device, or another option?

As you address these concerns, you may come up with a strategy that requires all user data to be backed up once a week to an on-premises NAS device. Alternatively, you may opt for hourly, automated backups to cloud storage services for mission-critical files and folders. For every scenario you can think of, a backup strategy can be developed to handle it.

Pro tip

One strategy used by systems administrators when planning backups is the **3-2-1 Rule**. This tried and tested method for ensuring data retention has three tenets: keep at least three copies of your data, store them on two different types of storage media, and ensure that one copy is kept offsite. By applying these rules, you can ensure that your data is recoverable in almost any failure scenario.

ChromeOS's approach to backups makes addressing these concerns even easier. As with other key security features of the operating system, backups occur somewhat naturally on ChromeOS just because of the way that the operating system is designed. For example, backing up user files isn't as critical on ChromeOS as it may be on other operating systems because by default much of your data is stored in the cloud.

That being said, even ChromeOS stores some data on local devices that could benefit from routine backups. Also, apps and custom configurations that are not the default or native to ChromeOS will need to be backed up in case there is a system failure.

To start things off, let's look at how we can back up our data to Google Drive.

Backing up to Google Drive

ChromeOS takes the complexity out of backups by handling much of the work automatically. By default, files, apps, and user settings are captured without you ever having to lift a finger. The one exception is **offline files**. These offline files are created when you enable the **Offline** feature in Google Drive and then access a document stored in Drive. With the feature enabled, you are given the option to download your files to your ChromeOS device's local storage. This enables you to work with the file without being connected to the internet.

To activate this feature, simply follow these steps:

1. Launch the Google Drive app and click on the settings gear.

2. In the **General** section of the **Settings** menu, scroll down to the **Offline** option as seen in *Figure 5.5*.

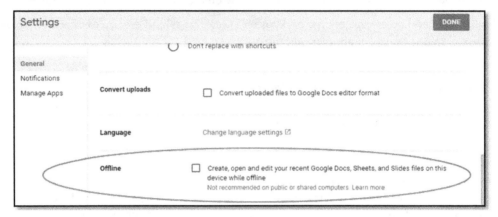

Figure 5.5 – Google Drive Offline setting

3. To enable the **Offline** option, check the checkbox to select it and then click the **DONE** button.

To back up your offline files following the activation of this feature, follow these steps:

1. In ChromeOS, navigate to **Settings → Device → Storage Management**.

2. From the **Storage Management** menu, select the **My files** option.

3. On the left side of the **My files** screen, you'll see the **Downloads** folder. Click on it to view its contents as illustrated in *Figure 5.6*.

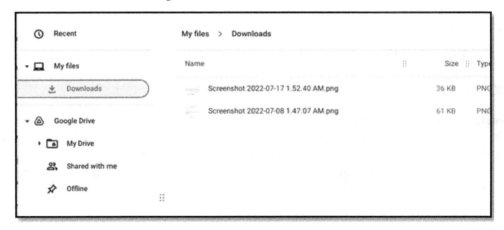

Figure 5.6 – My files à Downloads folder

4. To back up the files located under **Downloads**, drag them to the Google Drive option on the left side of the screen.

Note that if you have subfolders under the **My Drive** section of Google Drive, you can back up files to specific folders. You can also make these files available in offline mode by using the right-click menu options.

Now that you have a grasp of the *how* and *why* of backing up data to Google Drive, let's learn about what steps we should take to make backups to an external storage device.

Backing up to external storage

Backing up to external storage devices such as flash drives, SD cards, or external hard drives is a great way to ensure your data is recoverable and easily accessible. Unlike Google Drive backups, this method will allow you to physically secure additional copies of your data in an offline format.

Follow these steps to back up your data to an external storage device:

1. Connect your external storage device to your ChromeOS system.

    You may be asked to format the storage device the first time it is connected to the ChromeOS system. To do this, simply follow the onscreen instructions and wait for the process to complete.

2. In ChromeOS, navigate to **Settings** à **Device** à **Storage Management**.

3. From the **Storage Management** menu, select the **My files** option.

4. On the left side of the **My files** screen, you'll see the **Downloads** folder. Click on it to view its contents.

5. To back up the files located under **Downloads**, drag them to the external storage device displayed on the left side of the screen, as seen in *Figure 5.7*

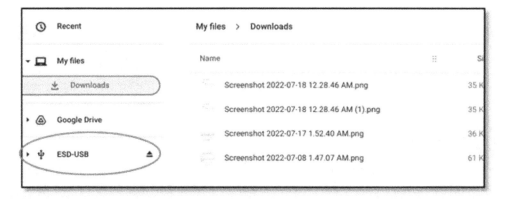

Figure 5.7 – External drive option in the My files section

Note that you'll be able to identify the external storage device based on the symbol next to the device's name. You'll see a USB symbol for flash drives/external hard drives and an icon of a storage card for SD cards.

There you have it! Your local, offline files are now recoverable in the event of a data disaster. In the next section, we will see how we can recover ChromeOS itself and its apps following a data disaster.

OS and app recovery

Up to this point, our focus has been on making sure that user data is kept safe from authorized access and recoverable following a disaster. But what about the OS itself? After all, don't you need a working operating system if you intend to access the data that you've secured? Of course you do. So, to guarantee that you get your ChromeOS device back up and running following a major OS failure, Google provides an easy-to-execute offline recovery method.

In this section, you will learn how to implement the USB-based method for recovering the Chrome operating system and its default apps.

When should you recover ChromeOS?

Before you make the decision to carry out a system recovery for ChromeOS, it's important to make absolutely sure that one is necessary. This is mainly due to the fact that the recovery process puts your data at rest at risk. During a system recovery, all of the contents of your ChromeOS device's hard drive will be permanently erased to make way for a new, clean copy of the OS. So, hopefully, you've been backing up those offline files!

Here are a few valid reasons for performing a recovery:

- Receiving the **ChromeOS is missing or damaged** error message
- Continued OS failures after multiple system reboots
- Exhaustion of other troubleshooting options
- Continued errors after performing a factory reset using the **Powerwash** feature

Now that you've confirmed that system recovery is necessary, let's get the process going by moving into recovery mode.

Recovery mode

Once you have established the need to perform a recovery, the next step is to put your ChromeOS device into **recovery mode**. Unlike Powerwash, which rolls back the Chrome operating system to its out-of-the-box state, recovery mode puts your device into a state that prepares it for the complete removal of the OS. To launch recovery mode, perform these steps:

1. Remove any external devices that are plugged into your ChromeOS device (for example, USB mice, thumb drives, and so on).

2. Use the appropriate method to put your ChromeOS device into recovery mode:

    -   **For some Chromebooks** – Press the power button while holding the *Esc* + *Refresh* buttons, as seen in *Figure 5.8*.

Figure 5.8 – Chromebook keyboard shortcut #1 for recovery mode

    -   For other Chromebook models – Press the power button while holding the *Esc* + *Maximize* buttons.

    -   For tablet-style Chromebook models – Press and hold the *Volume Up* + *Volume Down* + *Power buttons* for 10 seconds and then release them.

    -   For some Chromebox models – Use a thin object, such as a paper clip, to press and hold the device's **recovery button** while powering on the machine. Release the recovery button once you see output displayed on the screen.

Your device is now in recovery mode and ready to receive a fresh, clean copy of ChromeOS. At this point, you'll have two options for performing a reinstall of the OS – internet-based or USB-based. Let's review both methods.

USB-based OS recovery method

The USB-based recovery method provides a solid, offline way for getting a new copy of ChromeOS onto your computer. As an added bonus, this method will also produce reusable installation media. This may come in handy if you're administering more than one ChromeOS device, as it eliminates the somewhat lengthy (15- to 20-minute) process of resetting the bootable USB drive every time a device recovery is necessary.

Use the following steps to convert your USB drive into a ChromeOS installation drive:

1. Open the Chrome web browser and navigate to `https://chrome.google.com/webstore/detail/chromebook-recovery-utili/pocpnlppkickgojjlmhdmidojbmbodfm?authuser=0` to download **Chromebook Recovery Utility**.

Figure 5.9 – Chromebook Recovery Utility

Chromebook Recovery Utility is a browser extension that can be installed on any ChromeOS, Windows, or Mac device running the Google Chrome web browser. This means that any computer can be used to make your ChromeOS recover USB, not just a Chromebook or Chromebox.

2. Once the utility is installed, launch it by clicking the **extensions** icon in the top-right corner of your Chrome browser window.

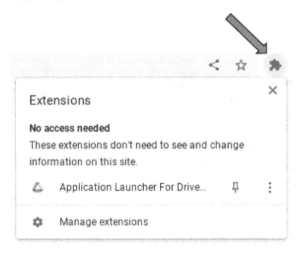

Figure 5.10 – Extensions icon

3. On the **Create a recovery media for your Chromebook** screen, click the **Get started** button, shown in *Figure 5.11*.

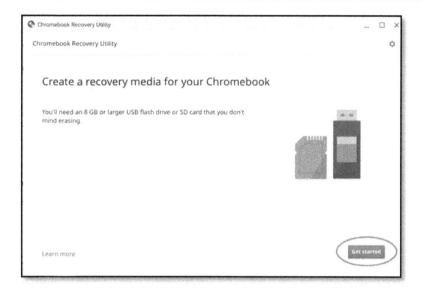

Figure 5.11 – The Create a recovery media for your Chromebook screen

4.   Next, on the **Identify your Chromebook** screen, either enter your Chromebook's model number manually or click the **Select a model from a list** option and choose your ChromeOS device's manufacturer and product name from the lists, as illustrated in *Figure 5.12*. Once your selections are made, click the **Continue** button.

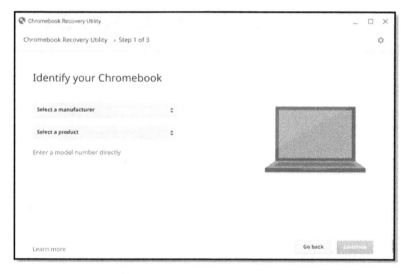

Figure 5.12 – The Identify your Chromebook screen

When the **Insert your USB flash drive or SD card** screen appears, insert your USB drive into the computer.

Once the drive is recognized, use the drop-down menu to select the USB drive as the one you'd like to use, as illustrated in the following screenshot, and then click the **Continue** button.

Figure 5.13 – The Insert your USB flash drive or SD card screen

5. On the **Create a recovery image** screen, click the **Create now** button, seen in *Figure 5.14*, to begin downloading ChromeOS to the USB drive. This process will take several minutes.

Figure 5.14 – Create a recovery image screen

6. When the USB drive creation has been completed, remove it from the working computer, insert it into the device experiencing the error, and select the **Recover using external disk** option.

Note that the device with your error message should already be in recovery mode based on the steps outlined previously.

7. Follow the onscreen instructions to complete the remaining system recovery steps.

8. Reboot the ChromeOS device to a working operating system.

With that, you now have yet another method of ensuring that your ChromeOS device can be returned to a usable state following a disaster.

Summary

We've come to the end of the chapter and you've gained a deeper understanding of how to protect and recover your data and system in ChromeOS. You received a crash course in the fundamentals of cryptography. We explored how ChromeOS hardware and software design utilizes encryption to protect our data automatically. You learned how to perform data backups of your offline files to Google Drive and external storage. Finally, you learned how to recover the entire OS and its default apps following a system failure.

In the next chapter, you'll continue to develop your troubleshooting expertise by learning how to fix common ChromeOS system issues.

6
Troubleshooting 101

So far, you've learned that the ChromeOS platform can be a very useful tool for many of your everyday and advanced computing tasks. The rock-solid performance it provides has made the operating system, and the devices that run it, increasingly popular among users and organizations alike. Yet even with all of its safety, security, and ease of use, it is still susceptible to failures from time to time.

In *Chapter 5*, *Recovering from Disasters*, you learned how to protect the data that your ChromeOS devices store and manage but what about the device itself? Your ChromeOS device, though typically lower in cost than other computing devices, still represents an investment. Therefore, to protect that investment, it is important to understand how to identify and correct issues with your ChromeOS device and software before they escalate into full-blown disasters.

In this chapter, you'll learn to troubleshoot the following:

- Crashing and freezing
- Slow system performance
- Constantly refreshing tabs
- No network connectivity
- Missing OS

So, are you ready to do some troubleshooting? If so, let the fun begin.

Technical requirements

In order to follow along with the activities outlined in this chapter, you'll need access to the following:

- A device with ChromeOS or ChromeOS Flex installed
- Wired or wireless internet
- A working computer or mobile device for online research

Crashing and freezing

System crashing and **freezing** are among the most annoying issues that you'll encounter on any computer system. Crashing is the sudden and abrupt shutting down of your computer system's operating system and hardware. Freezing is a locking up of your computer's software. In both cases, your system becomes unusable while the issue persists.

Even though crashing and freezing affect your system differently, both issues typically have the same root cause: corruption in your software code. Try the following options to fix the issue:

- **Reboot your system**:

  - Performing a restart of your computer system is a tried and true method of clearing corrupt software code from your system's RAM. This allows your machine to return to a clean state.

  - In the event that the normal reboot options aren't accessible (for example, your system is frozen), you can perform a hard reset of a ChromeOS device by pressing the keyboard combination *Ctrl + Shift + R*, as illustrated here:

Figure 6.1 – Hard reset keyboard shortcut

- **Remove recently installed apps and extensions**.

  If you're able to access your **Settings** menu, try removing any newly installed apps or extensions. The new code that they've added to your machine may be unintentionally, or even intentionally, causing instability on your ChromeOS device.

- **Perform a system reset**:

  - Performing a factory reset on the system will return ChromeOS to its out-of-the-box state. This will clear all of your information from the device's hard drive along with any corrupted files or software that could be the cause of the issue.

- Resets are performed by navigating to the **Settings** menu and selecting **Advanced → Reset settings → Powerwash** and then pressing the **Reset** button, as shown in *Figure 6.2*:

Figure 6.2 – ChromeOS Powerwash option

- **Perform a full system recovery**:

  - If all else fails, perform a system recovery on your ChromeOS devices. This will completely remove and replace the Chrome operating system as well as all apps and data on the system.

  - Detailed instructions for performing the system recovery process are discussed in *Chapter 5, Recovering from Disaster*.

Next, let's troubleshoot a slow-running Chrome device.

Slow system performance

The statement *my computer is running slow* is very common but also very ambiguous. What does *running slow* really mean? As the owner or administrator of a ChromeOS system, this is where you'll really need to utilize your detective skills to decipher the true meaning of slowness.

In most cases, one of two things is usually being referred to when a system is deemed slow:

- **Network speed**: How fast the user can surf the internet

- **Access/processing speed**: How long it takes the CPU to perform actions such as launching an app or opening a file

When dealing with network or internet-related speed issues, try these troubleshooting options to speed things up:

- **Close unused browser tabs**:

  Every web browser tab you launch consumes system resources such as RAM. The more you run, the fewer available resources you have, and the slower your system runs overall. This especially impacts your web-surfing capabilities

- **Reboot your system**:

  If closing your browser tabs doesn't free up enough resources to improve network performance, reboot the system to allow the computer to return to a clean state.

- **Check your Chrome device resource usage**:

  Just like most operating systems, ChromeOS provides a **Task Manager** utility that can be used to view the **processes** running on your machine. A process is any program that runs on your computer. They can be apps that run in the foreground (i.e., on your screen), such as Google Chrome, or in the background (i.e., without screen output), such as the ChromeOS Storage Service utility, which helps manage the storage of the device. The more programs you have running, the more system resources (e.g., RAM, CPU, and network bandwidth) they require. This can slow your machine down considerably. *Figure 6.3* provides an example of the type of information provided by Task Manager:

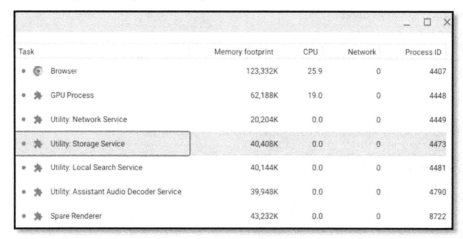

Task	Memory footprint	CPU	Network	Process ID
Browser	123,332K	25.9	0	4407
GPU Process	62,188K	19.0	0	4448
Utility: Network Service	20,204K	0.0	0	4449
Utility: Storage Service	40,408K	0.0	0	4473
Utility: Local Search Service	40,144K	0.0	0	4481
Utility: Assistant Audio Decoder Service	39,948K	0.0	0	4790
Spare Renderer	43,232K	0.0	0	8722

Figure 6.3 – ChromeOS Task Manager

You can access Task Manager in a few different ways. Some of the easier ways are by using the *Search + Esc* or *Launcher + Esc* keyboard shortcut or by clicking the Launcher and typing `Task manager` in the search bar. You can also access Task Manager by opening the Chrome browser, clicking on the three dots in the upper-right corner of the window (sometimes called the **kebab menu**), and navigating to **More tools** → **Task manager**, as seen in *Figure 6.4*:

Figure 6.4 – Accessing Task Manager via Chrome browser

Once you've accessed Task Manager, you can use it to identify any unnecessary processes or processes using too many resources and end them by selecting them and pressing the **End process** button, as seen in *Figure 6.5*:

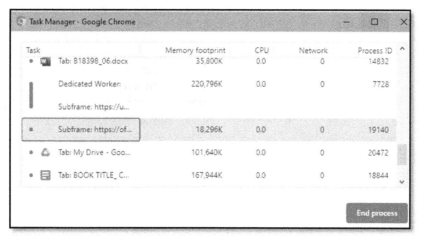

Figure 6.5 – End process option in Chrome Task Manager

- **Delete browser extensions**:

  In *Chapter 3*, *Exploring Google Apps*, we discussed how extensions can be used to add features and functionality to the Chrome browser. However, as with all application code, running extensions consumes system resources.

  To free up system resources and speed up the performance of your device, remove any unnecessary browser extensions by launching the Chrome browser, clicking the kebab menu, and then navigating to **More tools → Extensions**. There you'll see tiles representing each of the browser extensions installed on your ChromeOS device, as seen in *Figure 6.6*:

Figure 6.6 – Chrome browser extensions

  Each extension's tile will provide a toggle switch to enable or disable it as well as a **Remove** button. Performing either action will free up resources on your device, resulting in faster performance.

Pro tip

Even though browser extensions can add many enhancements to your Google Chrome user experience, they can also potentially introduce issues. Malicious software can sometimes masquerade as browser extensions. These corrupt extensions can be installed inadvertently as a part of app downloads from the Google Play Store. These malware browser extensions can track your web surfing habits and report them to a third party and even collect your personal information. Therefore, using the extension removal steps in this section not only will improve the performance of your device but they'll also increase your system's overall security.

- **Clear files from your local storage**:

  When storage begins to run low on any computer, it can cause performance issues. ChromeOS devices aren't immune to this hardware limitation. In fact, because most ChromeOS devices are designed to leverage Google Drive as their primary storage location for files, they tend to be limited to internal storage. When that storage fills up, performance slows down. So, try deleting some of those files you've downloaded or moving them to Google Drive to speed your system up.

- **Adjust your web service settings**:

For the sake of simplicity, ChromeOS devices don't overwhelm users with too many setting options. However as you have seen throughout this book, there are still a number of settings that users can configure and several of them can have a profound impact on the network performance of your device. By adjusting these settings, you can reduce your overall network activity, which will speed up your systems, especially when you have a weak or slower internet connection.

To modify your ChromeOS device's web service settings, navigate to the **Settings** menu and select **You and Google → Other Google services**, as illustrated in *Figure 6.7*. There, you will find options to disable **Autocomplete searches and URLs**, turn off **Enhanced spell check**, and stop URLs of pages you visit from being sent to Google:

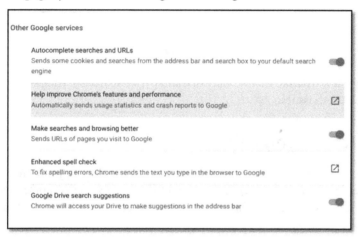

Figure 6.7 – Other Google services configuration screen

- **Change Google Drive settings**:

ChromeOS relies heavily on Google Drive to supply the majority of its storage services. As a result, more than just file uploads accumulate in Drive. All of your Gmail emails, docs, sheets, photos, and more call Drive their home. And just as traditional computers slow down as the hard drives fill up, so do our ChromeOS devices if **offline syncing** is enabled.

To disable Google Drive's offline sync functionality, launch the Drive app and click on the **Settings** gear. On the **Settings** page, uncheck the **Offline** option, as shown in *Figure 6.8*:

Figure 6.8 – Disabling offline sync in Google Drive

Now that you know a few tricks to resolve network and performance speed issues, let's move on to troubleshooting tab refresh issues.

Constantly refreshing tabs

ChromeOS devices can sometimes be resource-limited. No doubt by now you've realized that when those resources become exhausted, strange things begin to happen to your computer. One major issue that ChromeOS has inherited from the Chrome browser is the auto-reloading/refreshing of web pages when browser tabs are inactive.

In recent updates to the Chrome browser, some methods that were used to stop the auto-refresh/reload problem were *deprecated*. However, there are still a few troubleshooting processes that can be used to resolve the issue:

- **Turn off auto discarding**:

  Auto discarding is a feature that allows the Chrome browser to purge open and loaded browser tabs when a system is low on RAM.

  To turn this feature off, launch the Chrome browser and search for `chrome://discards`. Once there, you will see a list of active tabs and options to stop them from auto-reloading, as shown in *Figure 6.9*:

Discards		Database		Graph	
Loading State	**Lifecycle State**	**Discard Count**	**Auto Discardable**	**Last Active**	Acti
unloaded	discarded (urgent) at 12/16/2022, 2:15:04 AM	1	✓ [Toggle]	4 days ago	Load Disc
unloaded	discarded (urgent) at 12/16/2022, 2:15:19 AM	1	✓ [Toggle]	4 days ago	Load Disc Urg
loaded	hidden	0	✓ [Toggle]	4 days ago	Urge Disc
loaded	hidden	0	✓ [Toggle]	13 hours and 33 minutes ago	Load Urge Disc
				13 hours and	

Figure 6.9 – Partial output of the chrome://discards screen

Unchecking each tab's **Auto Discardable** option will force the browser refreshing to stop temporarily.

Unfortunately, this configuration only remains in place as long as the Chrome browser remains open. The moment the browsers are closed and relaunched, the settings will revert to their original state.

- **Disable Chrome browser extensions**:

Malicious or corrupted browser extensions can be a cause of auto-refreshing browser tabs. As mentioned previously, they can be removed by navigating to the **Extensions** menu and pressing the particular extension's **Remove** button.

- **Reset Chrome browser**:

Similar to the ChromeOS reset option, the Chrome browser reset restores the browser to its original state. This purges the browser of all extensions, cache, and history.

To perform this reset, launch the Chrome browser, click on the kebab menu, and select **Settings**. From there, select **Reset settings** → **Restore settings to their original defaults**, as shown in *Figure 6.10*:

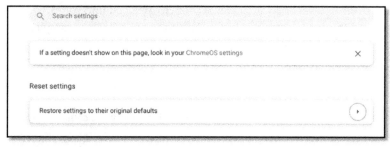

Figure 6.10 – Chrome browser restore settings option

Once the **Restore settings to their original defaults** option is selected, you will see the **Reset settings?** window, which will provide additional details on what the reset will accomplish. Once you have reviewed the additional information, click the **Reset settings** button, as shown in *Figure 6.11*:

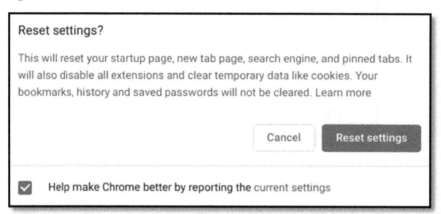

Figure 6.11 – Reset settings confirmation window

Our refreshing browser tab issue should now be a thing of the past! So now, let's learn how to handle network connectivity issues.

No network connectivity

As a cloud-first OS, internet connectivity is of the utmost importance to ChromeOS. Without a solid internet connection, many of the features and applications that make ChromeOS exceptional become unavailable. Therefore, administrators must have an understanding of how to resolve network connectivity issues regardless of the type of network their device is connecting to.

If you are having issues connecting to an **Ethernet network** (that is, wired internet connection), check the following:

- Ensure that your network cable is properly connected.

   Make sure that the Ethernet network cable is properly plugged into your ChromeOS Ethernet port or USB-to-Ethernet adapter and into your internet source (for example, router, network switch, network wall jack, et cetera).

   If properly connected, you will see the **Connected to Ethernet** message in the status tray.

- Try connecting a different device using the same Ethernet cable.

   This test can be used to rule out the network cable itself as being the cause of connectivity issues.

If you are having issues connecting to a **Wi-Fi network** check the following:

- **Ensure that Wi-Fi is enabled on your device.**

  Begin by opening the status tray and confirming the display of the name of the network you want to connect to. If the network isn't connected, you will see a **Not connected** message, as shown in this figure:

Figure 6.12 – Network not connected status

  To connect to a Wi-Fi network, simply click on the Wi-Fi fan icon, choose your wireless network from the list of network options that will display, and enter the Wi-Fi network password (if required).

- **Test the network from another device.**

  If you still can't connect to your Wi-Fi network, try connecting with another device to determine whether the network itself is the issue or whether it's your device. If the issue is determined to be device related, consult your ChromeOS device's manufacturer documentation for hardware-specific troubleshooting steps.

- **Make sure the Wi-Fi switch is turned on.**

  Some ChromeOS device models have a physical switch that allows you to toggle the Wi-Fi connections on and off. Review your device's hardware documentation to determine whether your computer has that feature, and then ensure that it is enabled.

- **Disconnect and reconnect to the network.**

  Sometimes, wireless network connectivity can be unreliable. In those instances, you may have a connection to a wireless network but not to the internet. If this occurs, try disconnecting from the wireless network and then reconnecting to reestablish your internet connectivity.

- **Run a ChromeOS connectivity diagnostic.**

  ChromeOS has a built-in, automated network diagnostic tool that will test your system's various forms of network connectivity. To access this tool, navigate to the **Settings** menu and then select **About ChromeOS → Diagnostics**, as shown in *Figure 6.13*:

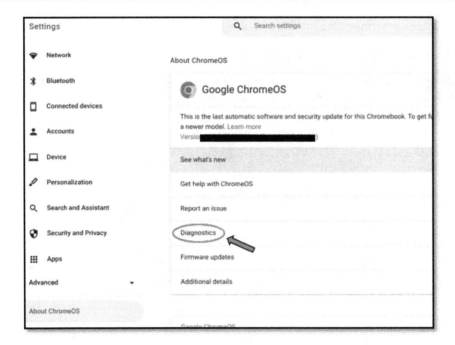

Figure 6.13 – ChromeOS Diagnostics tool

Once the **Diagnostics** window opens, select the **Connectivity** option. This will initiate an automated series of network tests and display the result of each test on your screen, as shown in *Figure 6.14*. The results of the report can aid you in repairing the detected network issue.

Figure 6.14 – Network connectivity diagnostics

If you are having issues connecting to a **mobile network**, check the following:

- **Ensure that mobile data is enabled**.

  As explained in *Chapter 2, Getting Connected*, some ChromeOS devices can connect to the cellular network. This allows them to use mobile data instead of wired or Wi-Fi connections.

  By navigating to **Settings → Network**, you can confirm that the **Mobile Data** network option is available and activated.

- **Ensure your mobile SIM card is properly installed**.

  Some ChromeOS devices are equipped with a SIM card slot that allows you to install a mobile carrier's SIM card directly into your computer. If the card is not properly seated in its tray, as shown in *Figure 6.15*, it can cause mobile network connectivity issues:

Figure 6.15 – Chromebook with built-in mobile SIM slot

After reseating the SIM, you may also need to open a command-line **Crosh terminal** on your ChromeOS device by pressing *Ctrl + Alt + T* and running the modem_set_carrier command followed by the name of your mobile carrier (for example, modem_set_carrier Verizon Wireless) and then press *Enter*. We'll discuss the use of the Crosh terminal in greater detail in *Chapter 8, Working with the Chrome Shell (Crosh)*.

So now that you have a better understanding of how network issues can be resolved in ChromeOS, let's learn what to do when your operating system becomes unavailable.

Missing OS

The Chrome operating system is at the core of your computer's overall functionality. So, it is of the utmost importance that it remains healthy and running. However, there are instances where the operating system experiences issues that prevent it from loading correctly.

When this happens, try the following options to resolve the issue:

- **Perform a full shutdown**:

  Sometimes, corruption in running software code can cause the OS to have issues. To resolve this problem, you can try unplugging all peripherals from your ChromeOS device and fully shutting the system down. When it powers back on, the corrupt code may be cleared and the issue resolved.

- **Perform an OS recovery**:

  If the OS issues persist following a shutdown and restart of your system, you may need to perform a full **OS recovery**. The process for performing the recovery is fully detailed in *Chapter 5, Recovering from Disasters*.

There you have it! You are now fully prepared to troubleshoot most of the major issues that are experienced by the ChromeOS.

Summary

In this chapter, you gained a solid understanding of the common issues experienced by the Chrome operating system and the steps needed to troubleshoot them. You learned how to overcome system freezing and crashing. You mastered tools and techniques for dealing with slow system performance and constant browser refreshing. Finally, you learned how to overcome network issues and operating system failure. These techniques will help you to keep the ChromeOS systems you support healthy and error-free so that you can continue to enjoy this powerful and flexible platform.

In the next chapter, we'll take our administration of the ChromeOS to the next level by fully tapping into the OS's Linux features and functionality.

Part 3: Advanced Administration

In this part, we will discover the advanced administration tools and techniques needed to take you from being a regular ChromeOS user to a full-fledged ChromeOS systems administrator.

This part comprises the following chapters:

- *Chapter 7, The Linux Development Environment*
- *Chapter 8, Working with the Chrome Shell (Crosh)*
- *Chapter 9, Google Workspace Admin Console*
- *Chapter 10, Centralized Administration of OUs, Users, Groups, and Devices*

7

The Linux Development Environment

By this point in this book, the capabilities of ChromeOS have been made more than evident. It provides many of the tools and features you'd expect from a widely adopted **operating system** (**OS**). Additionally, as the world's first commercially successfully Cloud-First operating system, it also boasts many features not present in traditional operating systems. However, in its native state, ChromeOS does have a major limitation: the inability to install and use full software applications and the tools used to develop them.

Thankfully, Google has overcome this limitation by enabling ChromeOS to tap into its Linux roots via its Linux development environment feature. This environment provides software developers using ChromeOS with a built-in Linux virtual machine, giving them an easily accessible and secure place to write code and develop apps.

In this chapter, you'll learn about the following topics:

- Enabling the Linux development environment
- Managing device permissions and Linux backup/restore
- Managing Linux storage
- Enabling Android Debug Bridge (ADB)
- Implementing port forwarding

So, let's unlock the next level of ChromeOS's power by unleashing its Linux capabilities.

Technical requirements

To follow along with the activities outlined in this chapter, you'll need access to the following:

- A device with ChromeOS that supports LDE. You can find out if your system can support LDE here: `https://sites.google.com/a/chromium.org/dev/chromium-os/chrome-os-systems-supporting-linux`.

- A wired or wireless internet connection.

- Additional internal or SD storage to accommodate Linux applications.

Enabling the Linux development environment

As you learned previously, ChromeOS is a modified version of the Linux operating system. However, many of the features typically left accessible in a normal Linux distribution have been omitted to increase the simplicity and security of the operating system, making it more user-friendly. The interface that you are presented with is a combination of the Chrome Desktop and browser but behind the scenes, Linux is still there hard at work. With a few minor configuration tweaks, you can unlock a host of additional applications, turning your ChromeOS system into a true multi-purpose computing device.

Let's get started by gaining a better understanding of what the Linux development environment is and what it does.

What is the Linux development environment?

In the early days of ChromeOS, administrators had to use third-party software hacks to access the Linux features of ChromeOS. However, now, you can easily unlock these built-in tools by enabling the **LDE** on your ChromeOS device. The LDE is a pre-built Linux virtual machine that runs as a computer within your computer. It requires very little configuration but gives you access to many of the **graphical user interface** (**GUI**) versions of Linux software applications.

Turning on the LDE

To enable the LDE on your ChromeOS device, follow these steps:

1. Navigate to the **Settings** menu and select the **Advanced** menu options.
2. Under the expanded **Advanced** menu, you will see the **Developers** menu option. Select it to view the **Linux development environment** option seen in *Figure 7.1*:

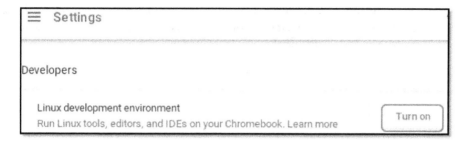

Figure 7.1 – Linux development environment option

3. Next, press the **Turn on** button to initiate the LDE activation process.

4. Once the initialization process begins, you'll be presented with the **Set up Linux development environment** screen. Here, you'll be informed of the benefits of enabling this feature and how much storage will be required to perform the installation. Click on the **Next** button to continue with the setup.

5. On the next screen, you'll be asked to create a **Username** to use when working on your Linux virtual machine. You'll also need to set the **Disk Size** option, which will indicate how much storage your Linux app will have access to use. The recommended amount of storage is 10 GB but this can be customized. Once you have made your configuration changes, click on the **Install** button, as illustrated in the following screenshot:

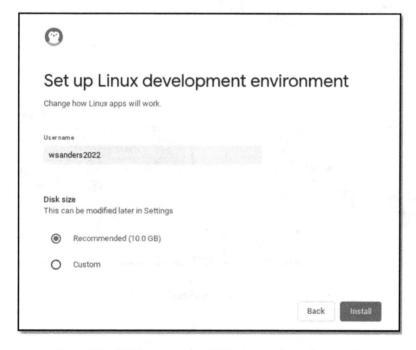

Figure 7.2 – LDE Username and Disk size configuration screen

6. After clicking the **Install** button, ChromeOS will begin downloading and installing the Linux VM onto your ChromeOS device. Once it completes this process, you will be presented with a Linux Terminal window, signaling that your installation was a success.

With the Linux development environment in place, you can now begin to look at some of the cool things it allows you to do.

Linux features

Installing the LDE opens up a whole new world of computing opportunities for ChromeOS users. As an administrator, you can now utilize command-line tools, code editors, and **integrated development environments (IDEs)** to code Linux apps. Additionally, now that your ChromeOS device is officially a Linux computer, you also have access to download and install thousands of Linux software applications, many of which are free!

So, how do you get started? Well, the easiest way to get these additional Linux tools onto your ChromeOS device is to find the *Linux* versions of the software you want on a vendor's website. Since Linux is a pretty popular OS platform, many software vendors create Linux-compatible versions of their applications. Simply check their website to see whether you can download the tool you need. For example, popular applications such as **Discord**, the voice/text/video messaging platform, provide Linux versions of its software for direct download from their website, as illustrated in *Figure 7.3*:

Figure 7.3 – Linux app download for Discord

Pro tip

In the Linux world, there are two major groups that the hundreds of Linux distributions out there fall into: **Red Hat-based** and **Debian-based**. Red Hat is an open source, commercial version of Linux that typically has a cost associated with the core OS and the applications that run on it. However, there are also free Red Hat-based OSes. **Red Hat Package Manager** is a component of the OS used to download and install software packages onto computers running Red Hat-based operating systems. Their software packages are easily identified by their .rpm file extension. Debian, on the other hand, is a completely free OS and all Debian-based OSes are also free. As a bonus, the software that runs on Debian-based systems is also typically free. Debian-based systems use the **Debian Package** (**DPKG**) manager utility to download and install software packages. The packages are easily identified by their .deb file extensions.

The Linux virtual machine that is deployed to ChromeOS when you enable the Linux development environment is a Debian-based version of Linux called **Gentoo**. This means that ChromeOS admins should look for .deb software packages when they want to download Linux apps.

Even though there will be software vendors who provide direct downloads of the Linux versions of their software from their websites, thousands of other applications and utilities have to be downloaded the old-fashioned way using the command-line Terminal.

To download Linux applications onto your ChromeOS device using the Terminal, follow these steps:

1. First, click on the **Launcher** area and locate the **Linux apps** container. In the container, you'll see the **Terminal** app, as illustrated in *Figure 7.4*:

Figure 7.4 – Linux Terminal app

2. When you launch the Terminal app, you'll be given several tools to choose from, as shown in
    the following screenshot:

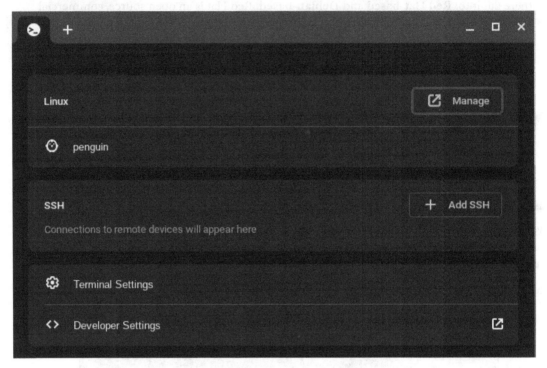

Figure 7.5 – Terminal menu options

The options include:

- **Linux**: This link is a shortcut to the **Advanced Settings | Developers** menu. With the LDE
  now in place, you'll see that there are several new configuration options available. We'll
  discuss these new options in detail later in this chapter.

- **Penguin**: This is a barebones, command-line-only installation of a Debian-based Linux
  system. It provides the *Terminal* that we'll use in the next few steps to install the software.

- **SSH**: This option gives you the ability to establish new or use existing **Secure Shell (SSH)**
  connections. SSH allows you to remotely control and administer servers.

- **Terminal Settings**: This option provides several configuration settings for adjusting the
  Terminal's text and background color, keyboard and mouse settings, system behavior, and more.

- **Developer Settings**: This is another link to the **Developer** menu.

3. Clicking on the **Penguin** option in the Terminal app will launch the actual command-line interface. At this point, you'll be presented with a dollar sign ($) command prompt.

4. To ensure that you're getting the most up-to-date versions of Linux apps, you must use the Terminal to run the `sudo apt-get update` command. This command causes your Linux VM to reach out to a Debian software repository to download the newest software packages available.

5. Once the general update has been completed, you can use the `sudo apt-get install <app name>` command to install the specific application you want to add to your ChromeOS system. In the following example, the `sudo apt-get install gimp` command has been used to install **GNU Image Manipulation Program (GIMP)**. GIMP is free software that allows you to edit photos and draws images digitally:

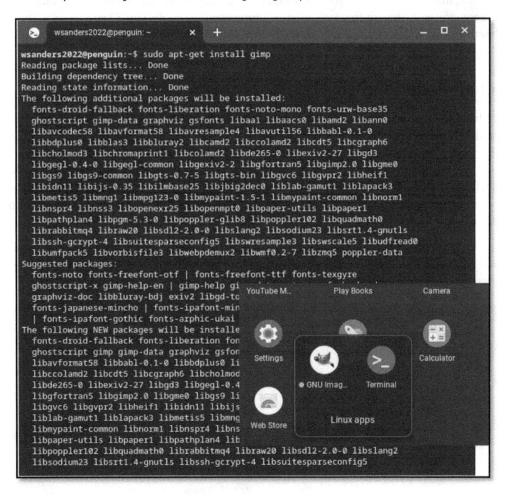

Figure 7.6 – Results of the Linux installation command for GIMP

Now that you have your Linux OS in place and your apps installed, let's explore how to keep them safe with permissions and a solid backup/restore process.

Managing device permissions and Linux backup/restore

As you learned in the previous section, the Linux development environment is a virtual Linux virtual machine running on top of ChromeOS. This means that many of the security measures discussed in *Chapter 4, ChromeOS Security*, do not apply to securing your LDE virtual machine. So, what does that mean for a ChromeOS systems administrator? It means that you're going to have to take the security concepts you've learned from managing ChromeOS's security and adopt them to the Linux operating system. Doing this will ensure that you can maintain a healthy Linux environment for projects and development efforts. Although this isn't a Linux administration guide, you'll learn exactly how to control permissions and perform data backups/restores using the tool integrated into ChromeOS to control your Linux environment.

Linux permissions

When addressing Linux permissions in ChromeOS, your focus will be primarily on controlling Linux's access to the ChromeOS device's hardware. The concept of sandboxing is the key to understanding how LDE permissions work. You may recall that sandboxing is critical to the security of ChromeOS. Each app and utility running on the OS or in the Chrome browser is isolated from one another. That way, if a failure or corruption occurs in one application, it won't affect the entire system. However, one drawback this presents to your LDE virtual machine is that, by default, it is isolated from accessing some of the hardware resources it may need to function.

To resolve this issue, ChromeOS includes several configuration options that allow you to grant Linux access to ChromeOS hardware that it may have otherwise been blocked from using. To configure it for this purpose, perform the following steps:

1. Navigate to the **Settings** menu and select **Advanced | Developers**.

2. When the **Linux development environment** option is enabled, you can expand the option to reveal several configuration settings, as illustrated in *Figure 7.7*:

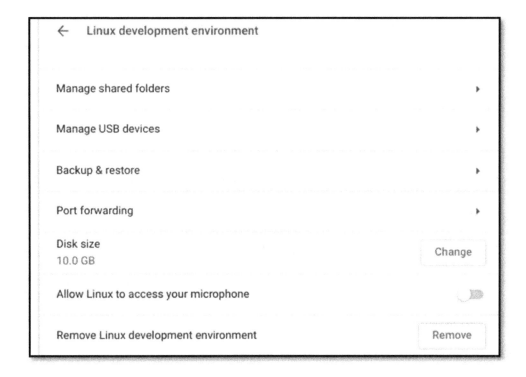

Figure 7.7 – Linux development environment configuration options

The options that control hardware permissions include:

- **Manage USB devices**: When selected, the **Manage USB devices** option will display a list of all of the supported USB devices that are currently plugged into your ChromeOS device. Each device will have a toggle switch beside it, allowing you to enable or disable Linux's ability to access the device. This option can be seen in *Figure 7.8*:

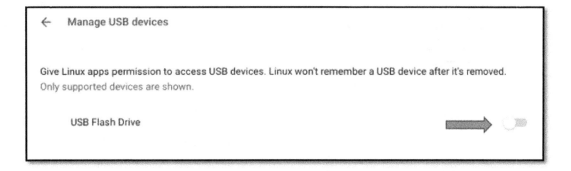

Figure 7.8 – The Manage USB devices option

- **Disk size – Change**: The **Disk size – Change** option allows you to adjust the amount of hard drive storage your Linux VM has access to. You configure this setting initially when you first enable the LDE. However, this option lets you adjust that setting to resize the system's virtual hard drive. This way, the system can be tailored to your storage needs as your usage of the LDE changes.

- **Allow Linux to access your microphone**: The **Allow Linux to access your microphone** option provides you with a simple toggle switch that can be used to enable or disable Linux's ability to use your system-embedded or USB-attached microphone for audio input.

With permissions taken care of, let's turn our attention to the Linux backup and restore process.

Linux backup/restore

We initially discussed the concepts of data backups and restores in *Chapter 5, Recovering from Disasters*. There, you learned the importance of backing up data in the event of system failures, accidental deletions, or any other event that may cause your data to become inaccessible. In the LDE, all of those same benefits still apply. However, the process of performing the backup/restore process is a little different.

To back up data from your LDE, perform the following steps:

1. Navigate to the **Settings** menu and select **Advanced | Developers**.
2. Under the **Developers** menu, click the arrow to open the **Linux development environment** options screen.
3. From the list of options, select the **Backup & restore** menu. Here, you'll be presented with buttons to initiate each of these tasks, as shown in the following screenshot :

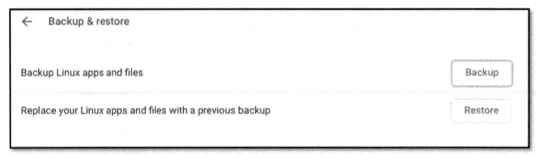

Figure 7.9 – The Backup & restore options screen

The following options are available:

- **Backup**: When the **Backup** button is selected, you will be redirected to the **Files** app, given the option to create a folder to save to, and prompted to name and save your backup file with a .tini file extension. The .tini file extension is a callback to the LDE's predecessor, **Crostini**, which was the Linux (beta) version of the tool:

Figure 7.10 – Linux backup file with the .tini file extension

- **Restore**: When the **Restore** button is selected, you'll be presented with a confirmation screen, as seen in *Figure 7.11*. This screen informs you that the restore will delete all of your existing Linux data and applications. After selecting **Restore**, you'll be taken to the **Files** app, where you'll choose the `.tini` backup file you want to use during the restore process:

Figure 7.11 – The Confirm Restore screen

> **Pro tip**
>
> If you run into issues performing the Linux backup, it may be due to storage size constraints on your ChromeOS device. The backup file that's created during this process can be rather large. Therefore, you may need to delete a few files from your system before trying again.

Now that we have the backup and restore process down, let's learn how to manage file and folder storage in the LDE.

Managing Linux storage

When setting up the LDE, you are required to set aside a portion of your ChromeOS device's storage to be used as the virtual hard drive for your Linux VM. Moving forward, this block of storage is used to house all of the apps, folders, and files that are produced by your Linux system. By default, these file and folder resources are only accessible to your Linux VM and its apps due to **sandboxing**. However, ChromeOS provides the **Manage shared folders** option to enable access to these Linux files and folders. When you initially select the **Manage shared folders** option, your screen will look similar to the following:

Manage shared folders

Shared folders are available in Linux at /mnt/chromeos. To share, right-click on a folder in Files app, then select "Share with Linux".

Shared folders will appear here

Figure 7.12 – The Managed shared folders option

The **Managed shared folders** screen briefly informs you that Linux shared folders are available in the /mnt/chromeos directory, which resides in your Linux VM. This is important to know in case you want to use the command line to access or move files around, or if you want a write a **shell script** that references files or folders stored in that directory. You'll learn more about scripting in *Chapter 8, Working with the Chrome Shell (Crosh)*.

Additionally, the **Manage shared folders** screen informs you that the folder-sharing process is initiated in the **Files** app, not through this tool.

To begin sharing folders and the files they contain, do the following:

1. Navigate to the **Settings** menu and select **Device | Storage management**.
2. Under the **Storage Management** menu, select **My files**.
3. When the **My files** screen opens, locate a directory or folder you want to grant access to and drag it to the **Linux files** menu option, which is made available when Linux development mode is enabled.

    Alternatively, you can right-click on a directory and select the **Share with Linux** option from the right-click menu. *Figure 7.13* shows both options:

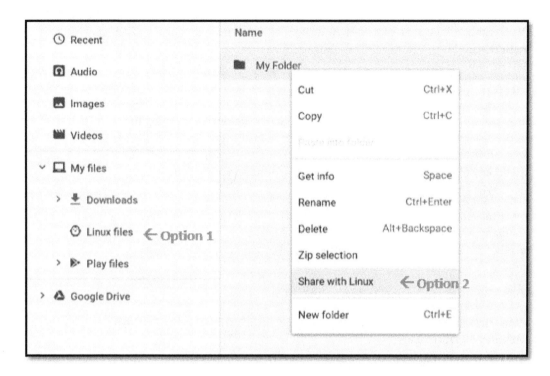

Figure 7.13 – Option for sharing files and folders with Linux

4. Now, navigate back to **Settings | Advanced | Developers | Linux development environment** and select the **Manage shared folder** options.

    You will now see the shared directories listed, along with an **X** button to stop sharing them, as seen in *Figure 7.14*:

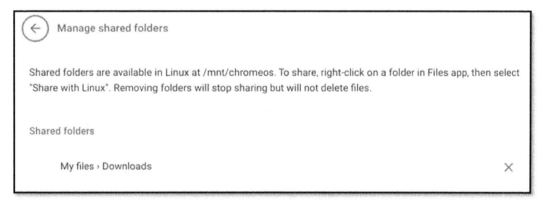

Figure 7.14 – Example of a folder shared with Linux

Now that you have a better understanding of how files and folders are shared between ChromeOS and its LDE, let's take a look at an awesome Android tool that is now at your disposal: **Android Debug Bridge (ADB)**.

Enabling Android Debug Bridge

ADB is a command-line tool that lets you communicate with devices running the Android operating system. It was created to help software developers code apps for Android devices so that they can interact with test devices for debugging purposes. With ABD, you can install and debug apps as well as run a variety of commands to perform various actions on your target device.

You may be thinking, "that sounds cool... but this is an Android tool. So, how can I use it?" Well, remember that, like ChromeOS, the Android OS is also derived from the Linux operating system. As a result, many Android apps and tools (including ADB) understand and respond to the Linux shell commands. This means that, with very little setup, you can have a fully functional test environment for Android apps at your disposal.

To set up ADB on your ChromeOS device, follow these steps:

1. With the Linux development environment enabled, navigate to the **Settings** menu and select **Advanced | Developers**:

2. Under the **Developers** menu, click the arrow to open the **Linux development environment** options screen.

3. Under the **Linux development environment** menu, select the **Develop Android apps** option, as seen in *Figure 7.15*:

Figure 7.15 – The Develop Android apps menu option

4. On the **Develop Android apps** screen, you are given a brief description of how ADB helps you to create test apps. You are also informed that, when enabled, the tool allows apps to be installed that haven't been verified by Google. To activate the tool, click the toggle switch next to the **Enable ADB debugging** option, as depicted in *Figure 7.16*:

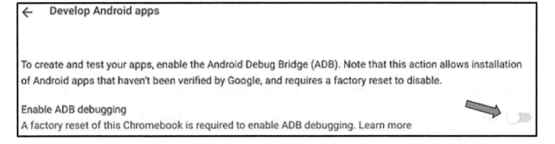

Figure 7.16 – Enable ADB debugging

5. Once the **Enable ADB debugging** option is switched on, you will be presented with a pop-up window, as shown in *Figure 7.17*, informing you that a restart of your system will be required to fully activate ADB debugging. You will also be warned that disabling the tool will require a factory reset (that is, a Powerwash) of the system. Click the **Restart and continue** button to finish activating ADB:

Figure 7.17 – Finalizing ADB activation

6. Once the **Restart and continue** button has been pressed, ChromeOS will shut down all running apps and restart your device. When the computer restarts, you'll once again see a message that warns you the device may contain apps that haven't been verified by Google. Confirm this warning and sign in to your ChromeOS device.

7. Once you're logged in, click on the **Launcher** button, navigate to **Linux app | Terminal**, and select the **penguin** option to open the command-line interface.

8. Finally, on the command line, type sudo apt-get install android-tools-adb to finalize the installation of ADB.

    Note that you may have to press **Y** to confirm and continue the installation.

With ADB in place, you can now begin using tools such as **Android Studio**, the official IDE for Android apps, to code and test apps from your ChromeOS device. For more information on Android Studio, visit the Android developer's website at https://developer.android.com/studio/intro.

It's important to note that not all ChromeOS devices are created equally. Therefore, setting up ADB may not be as straightforward, or even possible, in your computing scenario. So, if you don't see or can't activate the option for ADB, here are a few things you can check when troubleshooting:

- Ensure that you've allocated enough virtual disk space to your Linux VM as low storage can cause a myriad of issues on your VM.

- Ensure that incomplete configurations aren't blocking you from moving forward with ABD activation.

- Try restarting the system. In a worst-case scenario, you may need to perform a factory reset of the system to restart the configuration from the beginning.

If none of these options resolve the issue, your ChromeOS device likely does not support the direct use of ADB.

Now that you can test Android apps using ADB, let's move on to exploring this chapter's final Linux feature: port forwarding.

Implementing port forwarding

Port forwarding is another critical concept related to the development of apps on your ChromeOS device but to understand it, you'll need a short crash course in computer networking. In computer networking, a **port** is a number that has been associated with a **networking protocol** (that is, a set of rules for network communication) or an application. The numeric representations of protocols and apps are used as a kind of shorthand reference when defining how different applications will communicate over the network. For example, a web browser such as Google Chrome utilizes the **Hypertext Transfer Protocol Secure (HTTPS)** protocol to provide encrypted data communications over the internet. The HTTPS protocol is represented by port number 443.

When a device communicates over the network, it will indicate the port number of the protocol initiating communications (that is, the source protocol) and the port the communication is bound for on a destination computer (that is, the destination protocol). Networking devices, such as routers, are designed to recognize these port numbers so that they can direct the data packets that they are associated with to the proper application on the destination device. Your computer actively listens to these ports to receive and forward communications.

This process is important to ChromeOS users who leverage the LDE to develop Android apps because they may need to test how their apps will communicate over network connections. They may also need to connect their apps to external servers and devices using non-standard port numbers as part of the software's core functionality. In both instances, port forwarding on ChromeOS allows the local ports on your ChromeOS device to be opened and made accessible to external devices for communication.

To enable port forwarding in ChromeOS, perform the following steps:

1. With the LDE enabled, navigate to the **Setting** menu and select **Advanced | Developers**.

2. Under the **Developers** menu, click the arrow to open the **Linux development environment** options screen.

3. Under the **Linux development environment** menu, select the **Port Forwarding** option, as seen in *Figure 7.18*:

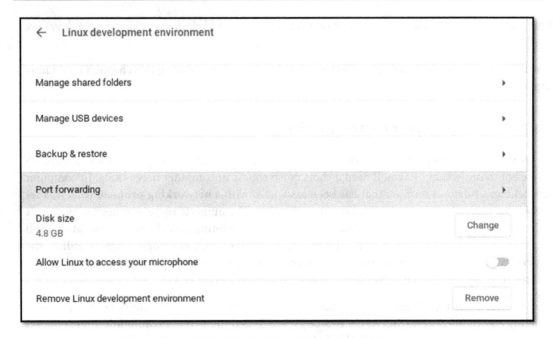

Figure 7.18 – The Port forwarding menu option

4. After selecting the **Port forwarding** menu option, you'll be presented with the option to add ports that you want to make available to Linux. By default, no ports are listed as being shared with Linux. This is illustrated in *Figure 7.19*. Click the **Add** button to begin forwarding ports:

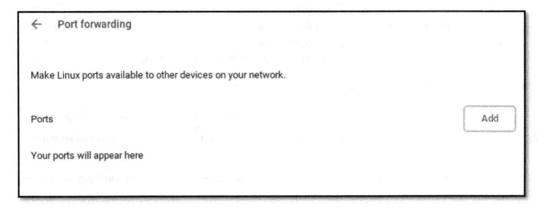

Figure 7.19 – The Port forwarding screen before ports have been forwarded

5. Once the **Add** button has been selected, you'll be presented with the **Add port number** configuration screen seen in *Figure 7.20*. Use it to configure your first forwarded port, and click the **Add** button once the fields are complete:

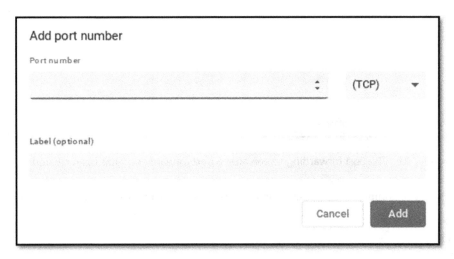

Figure 7.20 – The Add port number screen

The fields of the **Add port number** utility require the following information:

- **Port number**: This field requires you to enter the port number that you want to allow Linux to access. This port number needs to be between 1024 and 65535, to avoid overlap with a well-known port number already assigned to standard protocols.

- **TCP or UDP**: This drop-down menu needs to be set to the network transport protocol you've chosen to use to move data packets across the network, for the port you've chosen.

- **Label (optional)**: This optional field allows you to give the connection a name that is easy for you to recognize.

Note that you may need to use the **Add port number** utility several times if you want to create multiple port forward entries.

6. After creating the port number entry, you'll be taken back to the **Port forwarding** screen. You will now see the new port number you defined and its label (if one was configured) listed. This is illustrated in *Figure 7.21*. The entry will have a toggle switch so that the port can be disabled and re-enabled if necessary. Finally, the entry will also have a kebab menu, which is used to access the **Remove port** option:

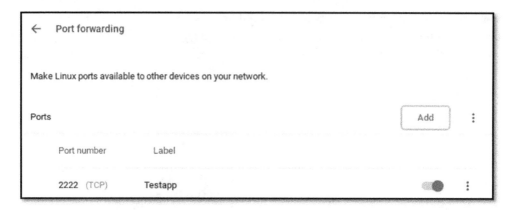

Figure 7.21 – The Port forwarding screen after a port forward entry has been created

And with that, we've completed our discussion on port forwarding and the major feature of the LDE.

Summary

The addition of the LDE to ChromeOS has taken an already powerful operating system platform to the next level. In this chapter, you learned how enabling the advanced functionality of Linux on top of ChromeOS can turn it into a device that advanced computer users can leverage to perform major technical tasks. You learned how to enable the LDE, access the Linux command-line Terminal, configure the LDE's permissions, and back up/restore its files. Additionally, you explored the process of managing Linux's access to system storage and enabling ADB and port forwarding for app testing and development.

In the next chapter, you will continue to build on our command-line experience by learning the essentials of the **Chrome Shell**, also known as **Crosh**.

8

Working with the Chrome Shell (Crosh)

In the previous chapter, you got a first glimpse of the command-line interface in ChromeOS via the Linux development environment's Terminal. The Terminal provides admins with the potential to access extremely powerful tools for managing and maintaining their computer systems and apps. However, at the end of the day, the Terminal is still on your system, primarily to help control the Linux VM running as a guest on your ChromeOS device.

Thankfully, Google has given ChromeOS a command line all of its own; **Chrome Shell** or as it's more commonly known, **Crosh**. Crosh can't be used to run arbitrary code like some other command-line interfaces. However, systems administrators and advanced users can also leverage this powerful, browser-based application to access powerful tools and utilities, many of which are not available in the GUI.

In this chapter, we will cover the following topics:

- Accessing Crosh
- Looking at the essential Crosh commands for systems administration
- Shell scripting

Technical requirements

In order to follow along with the activities outlined in this chapter, you'll need access to the following:

- A device with ChromeOS or ChromeOS Flex installed
- Wired or wireless internet

Accessing Crosh

Similar to the Windows Command Prompt and Linux bash, Crosh provides command-line capabilities for its operating system. Its main job is to provide access to the text-based tools and utilities needed to manage and maintain ChromeOS. However, unlike other OS command-line languages, Crosh is not actually built into the OS. This is one of the ways Crosh differs from Windows's command line and bash – it's browser-based, Chrome browser to be more precise.

To access Crosh on your ChromeOS device, simply press the *Ctrl + Alt + T* hotkey as illustrated in *Figure 8.1*. This will cause the Chrome browser app to launch a Crosh browser tab.

Figure 8.1 – Keyboard shortcut for Crosh

Unlike the Linux development environment's penguin terminal, Crosh has its own special set of commands. In fact, Crosh doesn't even recognize standard Linux bash commands because it has been purpose-built for ChromeOS system administration.

In the next section, we will take a look at some of the essential Crosh commands every administrator should know.

Looking at the essential Crosh commands for systems administration

Once you've launched Crosh, you'll be presented with a black screen and a command prompt reminiscent of the Linux terminal. If you're new to coding or the command line, this can be a bit intimidating, but don't worry – with the commands we'll cover in the section, you'll be *Croshing it* in no time (dad joke achieved)!

> **Pro tip**
>
> As an alternative to running Crosh from the browser, you can also download the **Crosh Window** app from the Chrome web store: https://chrome.google.com/webstore/detail/ crosh-window/nhbmpbdladcchdhkemlojfjdknjadhmh?hl=en-US.

help

When you don't know what to do next, the best thing you can do is ask for help. It's true in life and also at the command line. In the Crosh command prompt, you can use the help command to display a short list of commonly used commands, along with their descriptions, as seen in *Figure 8.2*:

```
crosh> help
exit
  Exit crosh.

help [command]
  Display general help, or details for a specific command.

help_advanced
  Display the help for more advanced commands, mainly used for debugging.

ping [-4] [-6] [-c count] [-i interval] [-n] [-s packetsize] [-W waittime] <destination>
  Send ICMP ECHO_REQUEST packets to a network host.  If <destination> is "gw"
  then the next hop gateway for the default route is used.
  Default is to use IPv4 [-4] rather than IPv6 [-6] addresses.

top
  Run top.
```

Figure 8.2 – Crosh help command output

help_advanced

Although the help command is, well, helpful, it may not provide you with the command you need to get the job done. In those instances, you can use the help_advanced command to display a more in-depth list of the commands Crosh has to offer.

uptime

Use the uptime command to determine how long your ChromeOS device has been powered on, as well as to obtain other statistics related to its operating time, as illustrated in *Figure 8.3*.

```
crosh> uptime
 00:27:31 up 10 days,  4:42,  0 users,  load average: 0.54, 0.48, 0.49
```

Figure 8.3 – Crosh uptime command output

top

The Chrome browser provides us with a GUI-based task manager that can be accessed by clicking on the kebab stack (three dots in the upper-right corner of the browser) and navigating to **More tools** à **Task Manager**. This tool shows you all of the running processes on your system and allows you to stop them if they are consuming too much of the system's resources. Alternatively, you can use the top command to view running processes directly from the command line.

Although top won't allow you to end processes as Task Manager will, it does reveal hidden, low-level processes that are invisible in the GUI version of the tool. *Figure 8.4* show a side-by-side comparison of these two process management utilities.

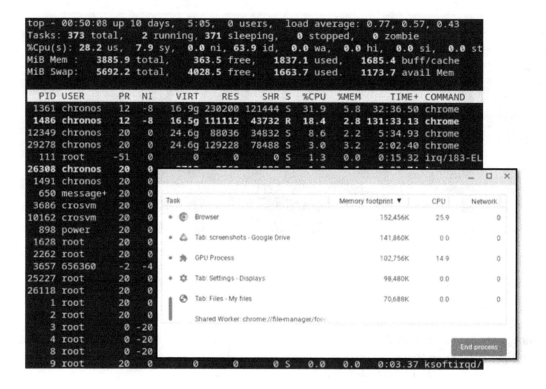

Figure 8.4 – Task Manager versus the top command output

Pro tip

Some Crosh commands, such as top, will continue to run indefinitely until they're forced to terminate. To force a command to stop at the Crosh command line, you can use the *Ctrl + C* hotkey. Additionally, since there is no command in Crosh to clear command output from your screen, just refresh the browser tab to clear the prompt.

enroll_status

Your ChromeOS device has the ability to function as a standalone computer or a centrally managed workstation. The process of configuring a ChromeOS device to be part of an enterprise network is called **enrollment**. We'll explore this concept more in-depth in *Chapter 9, Google Workspace Admin Console*, but for now, know that the enroll_status command can be used to see whether your device is a part of an enterprise or not as seen in *Figure 8.5*.

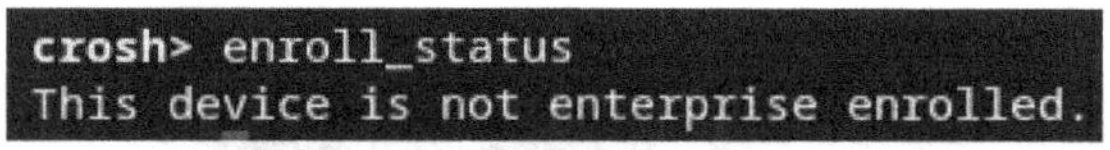

```
crosh> enroll_status
This device is not enterprise enrolled.
```

Figure 8.5 – Crosh enroll_status command output

ipaddrs

To determine the IP address settings on your ChromeOS device, you can use the ipaddrs command. There are two versions of this particular command, ipaddrs -4 and ipaddrs -6. The ipaddrs -4 command will display a system's IP version 4 configuration and ipaddrs -6 will display IP version 6 settings. *Figure 8.6* shows an example of the ipaddrs -4 command output.

```
crosh> ipaddrs -4
/0 [ ip -4 addr show ]
/1 1: lo: <LOOPBACK,UP,LOWER_UP> mtu 65536 qdisc noqueue state UNKNOWN group default qlen 1000
/2     inet 127.0.0.1/8 scope host lo
/3        valid_lft forever preferred_lft forever
/4 2: wlan0: <BROADCAST,MULTICAST,ALLMULTI,UP,LOWER_UP> mtu 1500 qdisc noqueue state UP group d
/5     inet 192.168.          global wlan0
/6        valid_lft forever preferred_lft forever
/7 3: arc_ns0@if2: <BROADCAST,ALLMULTI,UP,LOWER_UP> mtu 1500 qdisc noqueue state UP group defau
/8     inet 100.11              scope global arc_ns0
/9        valid_lft forever preferred_lft forever
/10 4: arc_ns1@if2: <BROADCAST,ALLMULTI,UP,LOWER_UP> mtu 1500 qdisc noqueue state UP group defa
/11     inet 100.1               scope global arc_ns1
/12        valid_lft forever preferred_lft forever
/13 6: arcbr0: <BROADCAST,MULTICAST,UP,LOWER_UP> mtu 1500 qdisc noqueue state UP group default
/14     inet 100.11             pe global arcbr0
/15        valid_lft forever preferred_lft forever
/16 7: arc_wlan0: <BROADCAST,MULTICAST,ALLMULTI,UP,LOWER_UP> mtu 1500 qdisc noqueue state UP gr
/17     inet 100.11            ope global arc_wlan0
/18        valid_lft forever preferred_lft forever
/19 9: arc_ns2@if2: <BROADCAST,ALLMULTI,UP,LOWER_UP> mtu 1500 qdisc noqueue state UP group defa
/20     inet 100.11             scope global arc_ns2
/21        valid_lft forever preferred_lft forever
/22 11: vmtap0: <BROADCAST,MULTICAST,ALLMULTI,UP,LOWER_UP> mtu 1500 qdisc pfifo_fast state UP g
/23     inet 100.11             ope global vmtap0
/24        valid_lft forever preferred_lft forever
```

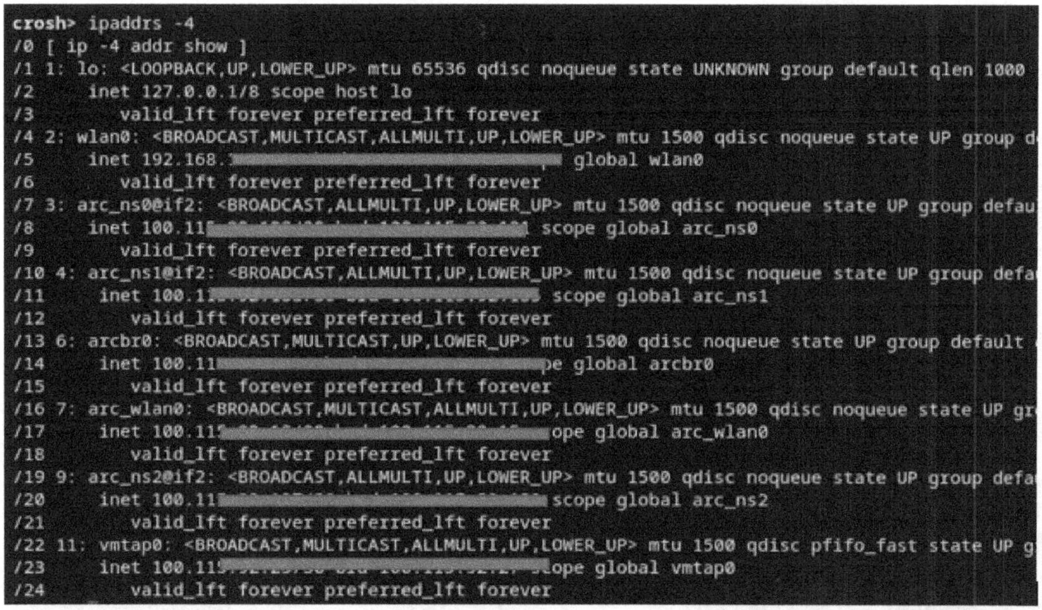

Figure 8.6 – Crosh ipaddrs -4 command output

ping

As with most other operating systems, the super popular `ping` command serves as the go-to command-line tool for performing connectivity testing. Unlike the other commands that we've discussed up to this point, the `ping` command's syntax is a little different. For `ping` to function properly, you have to type the `ping` keyword along with the IP address or domain name of the device whose connection you want to test. In the example here, `ping` is used to see whether `www.google.com` is up and running. Surprise! Google is doing just fine:

```
crosh> ping www.google.com
PING www.google.com (142.251.35.164) 56(84) bytes of data.
64 bytes from lga25s78-in-f4.1e100.net (142.251.35.164): icmp_seq=2 ttl=119 time=11.9 ms
64 bytes from lga25s78-in-f4.1e100.net (142.251.35.164): icmp_seq=3 ttl=119 time=10.3 ms
64 bytes from lga25s78-in-f4.1e100.net (142.251.35.164): icmp_seq=4 ttl=119 time=10.8 ms
64 bytes from lga25s78-in-f4.1e100.net (142.251.35.164): icmp_seq=5 ttl=119 time=11.7 ms
64 bytes from lga25s78-in-f4.1e100.net (142.251.35.164): icmp_seq=6 ttl=119 time=10.2 ms
64 bytes from lga25s78-in-f4.1e100.net (142.251.35.164): icmp_seq=7 ttl=119 time=12.8 ms
64 bytes from lga25s78-in-f4.1e100.net (142.251.35.164): icmp_seq=8 ttl=119 time=11.5 ms
64 bytes from lga25s78-in-f4.1e100.net (142.251.35.164): icmp_seq=9 ttl=119 time=11.8 ms
```

Figure 8.7 – Crosh ping command output

modem

If your ChromeOS device needs to leverage a modem for internet connectivity, the `modem` command provides several configuration options. Type the `modem help` command to see a full listing of its options as illustrated in *Figure 8.8*:

```
crosh> modem help
Usage: modem <command> [args...]
  activate [-modem <modem>] [<carrier>]           Activate modem
  activate-manual [-modem <modem>] [args...]      Activate modem manually
  ciprl-update                                    Enable and perform client initiated PRL update
  connect [-modem <modem>] [phone number]         Connect modem
  factory-reset [-modem <modem>] [<spc>]          Factory-reset the modem
  get-oma-status                                  Current OMA-DM setting
  get-prl                                         Current PRL
  get-service                                     Integer index of data service
  reset [-modem <modem>]                          Reset the modem
  set-carrier [-modem <modem>] <carrier-name>     Set modem carrier firmware
  set-logging (debug|info|warn|error)             Set logging level
  set-madison-config <config-group-name>          Set Madison config group
  start-oma                                       Enable and launch OMA session
  status                                          Display modem status
  update-prl [-modem <modem>] <prl-file-name>     Install a PRL file
  ussd [-modem <modem>] status                    Show status of ongoing USSD session
  ussd [-modem <modem>] initiate <command>        Initiate a USSD session
  ussd [-modem <modem>] respond <response>        Respond to a USSD request
  ussd [-modem <modem>] cancel                    Cancel ongoing USSD session
crosh>
```

Figure 8.8 – Crosh modem help command output

set_time

The set_time command allows you to manually configure the system time of your ChromeOS device. As with ping, this command requires an additional **time string** command-line **argument** in order to work. The format of the time string uses the **GNU core utilities (coreutils)** date command format, which is DD <month_name> YYYY HH:MM <am_or_pm>:

- DD = a 2-digit day (for example, 12)

- <month_name> = the name of the month completely spelled out (for example, August)

- YYYY = a 4-digit year (for example, 2022)

- HH:MM = a 2-digit hour and 2-digit minute (for example, 02:30)

- <am_or_pm> = either an a.m. or p.m. time setting

The result will look like this: set_time 12 August 2022 2:30pm.

> **Important note**
> Note that this command may not work on all devices, as some ChromeOS devices exclusively pull their time setting from a network time server.

battery_test

If you're curious to find out how the battery is holding up in your ChromeOS device, using the battery_test command will be a big help. The battery_test command displays not only the current level of charge in your battery but also the battery's health percentage, along with its electrical discharge rate. This is important because these can be key indicators as to when your battery may need to be replaced. An example of the command's output can be seen in *Figure 8.9*. Note that this command can be used without options but the administrator can also define the number of seconds the test should take (the default is 300 seconds).

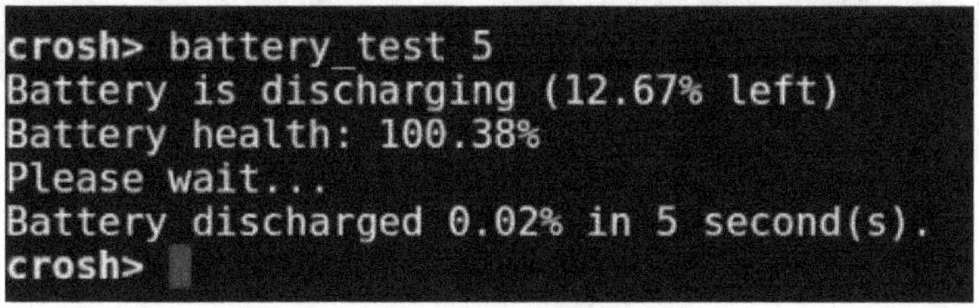

```
crosh> battery_test 5
Battery is discharging (12.67% left)
Battery health: 100.38%
Please wait...
Battery discharged 0.02% in 5 second(s).
crosh>
```

Figure 8.9 – Crosh battery_test command output

battery_firmware info

If you determine that your battery needs to be replaced, it would be helpful to have some information on its specs, right? Thankfully, Crosh also provides us with the `battery_firmware info` command. As you can see from the example here, this command displays important information regarding the make, model, chemistry, and so on of your battery:

```
crosh> battery_firmware info
EC result 3 (INVALID_PARAM)
Battery info:
  OEM name:             BYD
  Model number:         DELL FY
  Chemistry   :         LION
  Serial number:        0796
  Design capacity:      3684 mAh
  Last full charge:     2882 mAh
  Design output voltage 11400 mV
  Cycle count           137
  Present voltage       12786 mV
  Present current       0 mA
  Remaining capacity    2882 mAh
  Flags                 0x0b AC_PRESENT BATT_PRESENT CHARGING
```

Figure 8.10 – Crosh battery_firmware info command output

free

If you want to determine the memory utilization of your ChromeOS device, the `free` command has you covered. This command will provide you with a basic table summarizing your total amount of RAM, how much is currently in use for various purposes, and how much is available. Additionally, the command displays the **Swap** (virtual memory) stats for your view. *Figure 8.11* provides an example of the command output.

```
crosh> free
           total        used        free      shared  buff/cache   available
Mem:     3979160     1936432      336952      528168     1705776     1075248
Swap:    5828844     1649988     4178856
```

Figure 8.11 – Crosh free command output

meminfo

The `meminfo` command expands on the information provided by the `free` command. It offers a granular level view of a system's memory utilization, which may be helpful for advanced administrators and software developers. *Figure 8.12* shows a portion of the data provided by this command.

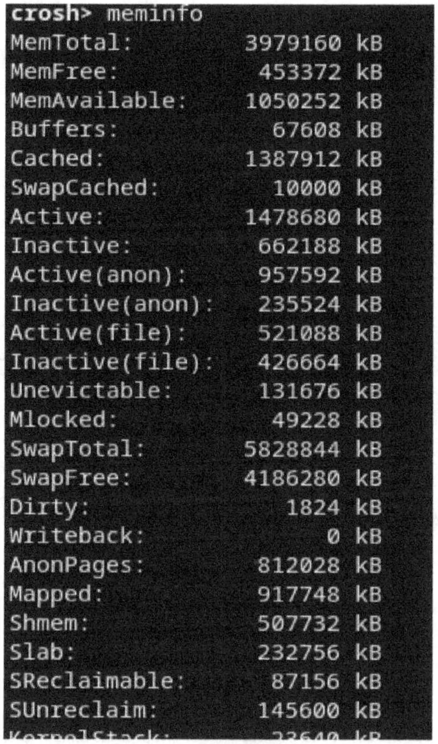

```
crosh> meminfo
MemTotal:        3979160 kB
MemFree:          453372 kB
MemAvailable:    1050252 kB
Buffers:           67608 kB
Cached:          1387912 kB
SwapCached:        10000 kB
Active:          1478680 kB
Inactive:         662188 kB
Active(anon):     957592 kB
Inactive(anon):   235524 kB
Active(file):     521088 kB
Inactive(file):   426664 kB
Unevictable:      131676 kB
Mlocked:           49228 kB
SwapTotal:       5828844 kB
SwapFree:        4186280 kB
Dirty:              1824 kB
Writeback:             0 kB
AnonPages:        812028 kB
Mapped:           917748 kB
Shmem:            507732 kB
Slab:             232756 kB
SReclaimable:      87156 kB
SUnreclaim:       145600 kB
KernelStack:       23640 kB
```

Figure 8.12 – Crosh meminfo command output

memeory_test

The memory_test command provides you with a command-line alternative to the GUI-based memory diagnostic tool located in the **Settings** menu under **About ChromeOS → Diagnostics → Memory**. A major difference between the command-line and GUI versions of this tool is the level of detail in the output. The GUI merely gives you a Pass or Fail output, while the command line provides an itemized list of every test memory performed. The figure here illustrates a side-by-side comparison of each of the tool's output.

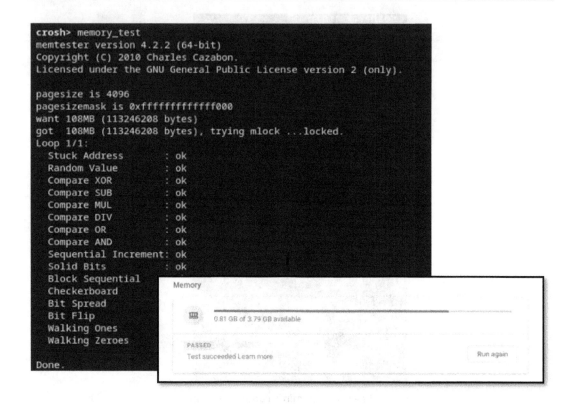

Figure 8.13 – Crosh memory_test command versus the Memory diagnostic tool

Ssorage_test

The storage_test command has two versions: storage_test_1 and storage_test_2. As you've probably guessed, each of these commands performs a diagnostic test on your ChromeOS device's storage. However, storage_test_1 performs a shorter, offline, **Self-Monitoring Analysis and Reporting (SMART)** test. In contrast, storage_test_2 provides an extensive readability test to ensure there are no bad blocks of storage. The output of storage_test_2 can be seen here in *Figure 8.14.*

```
crosh> storage_test_2
Checking blocks 0 to 30535679
Checking for bad blocks (read-only test): done
Pass completed, 0 bad blocks found. (0/0/0 errors)
```

Figure 8.14 – Crosh storage_test_2 command output

> **Pro tip**
>
> `storage_test_1` or `storage_test_2` may not be able to run for ChromeOS devices based on the type of hard drive you have. The `storage_test_1` command only works on systems that have **SSD**- or **NVMe**-style hard drives. The `storage_test_2` command only runs on systems with **eMMC**-style drives. Fortunately, you can test both commands on your system without risking any damage to your stored data. Also, if you discover a large number of bad storage blocks during either diagnostic test, it may be a good time to back up your data and perform a powerwash of the system.

network_diag

In the event that you encounter network issues on your ChromeOS device, you can use the `network_diag` command to perform a diagnostic test to attempt to pinpoint the source. Additionally, the results of the test are stored in a text file on your system for later review. The output of the `network_diag` command can be seen here:

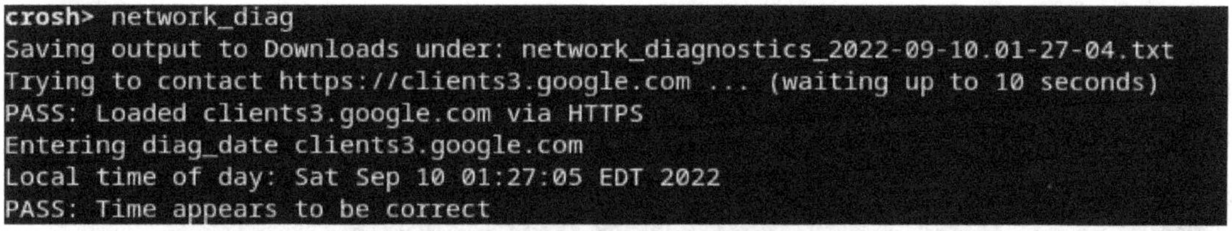

```
crosh> network_diag
Saving output to Downloads under: network_diagnostics_2022-09-10.01-27-04.txt
Trying to contact https://clients3.google.com ... (waiting up to 10 seconds)
PASS: Loaded clients3.google.com via HTTPS
Entering diag_date clients3.google.com
Local time of day: Sat Sep 10 01:27:05 EDT 2022
PASS: Time appears to be correct
```

Figure 8.15 – Crosh network_diag command

tracepath

As you continue your network troubleshooting, you may determine that your connectivity issue isn't caused by your system or network. Instead, the bottleneck may be occurring somewhere between your network and the remote network you want to communicate with. In that case, the `tracepath` command is the perfect tool for the job. This command traces the route that your data takes as it leaves your network and is bound for a remote network.

This command requires that you enter the domain name or IP address of the remote network you wish to reach for the command to function correctly. An example of the command's execution is shown here:

```
crosh> tracepath packt.com
 1?: [LOCALHOST]                          pmtu 1500
 1:  Fios_Quantum_Gateway.fios-router.home            2.497ms
 1:  Fios_Quantum_Gateway.fios-router.home            2.045ms
 2:  lo0-100.BLTMMD-VFTTP-309.verizon-gni.net         3.899ms
 3:  ae1309-21.TWSNMDTW-MSE01-AA-IE1.verizon-gni.net  6.773ms
 4:  no reply
 5:  152.                                            20.200ms
 6:  199.                                            13.427ms
```

Figure 8.16 – Crosh tracepath command output

route

When managing network communications on your device, you may want to see what communication connections have been established. Maybe you need to determine which network ports are open so that you can shut them down as part of your system hardening procedures. In either case, running the route command will be helpful. The route command allows you to view your system's **routing table**, the database that stores the route information that your computer learns as it communicates with remote networks. *Figure 8.17* shows a portion of the information that can be viewed by using route:

```
crosh> route
/0 [ ip -4 rule list ]
/1 0:    from all lookup local
/2 9:    from all lookup main
/3 10:   from all fwmark 0x3ea0000/0xffff0000 lookup 1002
/4 10:   from all oif wlan0 lookup 1002
/5 10:   from 192.1        /24 lookup 1002
/6 10:   from all iif wlan0 lookup 1002
/7 32765:      from all lookup 1002
/8 32766:      from all lookup main
/9 32767:      from all lookup default
/10
/11 [ ip -4 route show table main ]
/12 100.                dev arcbr0 proto kernel scope link src 100.
/13 100.                dev arc_wlan0 proto kernel scope link src 100.
/14 100.                dev vmtap0 proto kernel scope link src 100.
/15 100.                via 100.1           arc_ns0
/16 100.                dev arc_ns0 proto kernel scope link src 100.
/17 100.                via 100.            v arc_ns1
/18 100.                dev arc_ns1 proto kernel scope link src 100.
/19 100.                via 100.1           arc_ns2
/20 100.                dev arc_ns2 proto kernel scope link src 100.
/21 100.                via 100.            vmtap0
/22 192.                  wlan0 proto kernel scope link src 192.
/23
```

Figure 8.17 – Crosh route command output

rollback

In *Chapter 5, Recovering from Disasters*, you learned that sometimes the only way to get your ChromeOS system to a stable state is to roll back your recent changes. This is exactly what the `rollback` command does. With a single command, your system will attempt to revert to the previous update cached on the device.

vmc

Even though Crosh cannot directly execute Linux bash commands, it can forward commands to the built-in Linux VM. Two commands that take advantage of this functionality are `vmc start termina` and `vmc stop termina`. When the Linux development environment is enabled, the `vmc start termina` command allows you to open a terminal connection to your Linux container VM from the Crosh command line as illustrated in *Figure 8.18*. With the connection established, you can now run Linux commands.

```
crosh> vmc start termina
(termina) chronos@localhost ~ $ ls -al
total 8
drwxr-xr-x 2 chronos root 4096 Jul 17 02:31 .
drwxr-xr-x 3 root    root 4096 Jul 17 02:31 ..
```

Figure 8.18 – The vmc start termina command output and a sample of the Linux code run in Crosh

When you've completed running your Linux command, you can use the bash `exit` command to close the terminal and return to your Crosh command line. However, if you need to shut down the entire Linux container VM and its app, you can use the `vmc stop termina` command.

exit

Last but not least, we have the `exit` command. As the name implies, this command exits the Crosh command line by closing your browser tab. At first glance, this may not seem like a particularly useful command. However, as we begin to automate our systems administration tasks in ChromeOS, being able to programmatically exit the command line becomes essential.

Speaking of automation, in our next section, we're going build on our newfound Crosh command-line knowledge by learning to weave commands together into shell scripts.

Shell scripting

In the previous section, you were able to gain an understanding of how Crosh can be used to execute individual commands at the command line. However, one of the truly powerful features of any command-line language is the ability to take those individual commands and link them together into scripts. Scripts provide two important benefits to administrators; they allow logic to be introduced into the command line and they open the door to automation.

Due to its limited command set and the targeted nature of the Crosh commands, shell scripting can't be natively performed. However, as you've previously experienced, Crosh can be used as a pass-through to access the Linux command line hidden underneath. However, to leverage this feature, you'll need to make a few modifications to your system. Performing the following steps will prepare the way for you to do shell scripting on ChromeOS:

1. On your Chrome device's keyboard, press and hold the *Esc + Refresh + Power* keys as illustrated in *Figure 8.19*. This will cause the machine to reboot into **Recovery** mode.

Figure 8.19 – Keyboard shortcut to enter ChromeOS Recovery Mode

Note that proceeding forward from this point will clear all of the stored data that you have from your local system, so make sure you have a backup!

2. Once the **Recovery** mode screen has opened, press *Ctrl + D* to enter **Developer** mode. Developer mode is a configuration that gives you true root-level administrative access to your ChromeOS device. This mode is only meant for advanced users because it disables many of the protection mechanisms we discussed in earlier chapters. Once developer mode is triggered, you'll see the **Preparing System for Developer Mode** screen illustrated here:

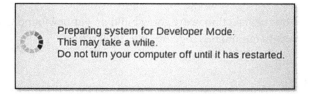

Figure 8.20 – The Preparing system for Developer Mode screen

3. After Developer mode completes its setup process, you'll be returned to the **Recovery mode** screen. However, the system will now indicate that **OS verification is OFF** as seen in *Figure 8.21*. This is evidence that your transition into Developer Mode was a success. From here, you'll press *Enter* to trigger a reboot of the system.

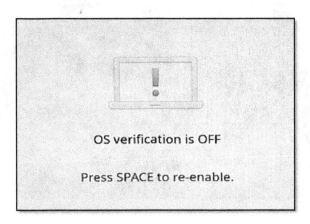

Figure 8.21 – Recovery Mode screen with OS verification disabled

4. Once the system reboots, you'll need to go through the initial system setup process outlined in *Chapter 1*, *ChromeOS Basics*. The good news is that your Google account will be able to reset any account personalization you implemented once you sign in.

5. With your basic setup out of the way, you can now press *Ctrl + Alt + T* to launch the Crosh command line.

6. In Crosh, you'll now run the newly enabled Shell command. Similar to the vmc start termina command, Shell allows you access to a Linux terminal from the Crosh. The huge difference is this ChromeOS terminal is the actual Linux core of ChromeOS and not just the Linux VM that's activated when you enable the Linux development environment.

7. With the Shell command launched, you can now begin using the full array of Linux commands on your Crosh browser tab. You can also launch command-line text editors such as **vim**, which allow you to create scripts without ever leaving the command line. Just be sure to store the script in a location on the filesystem that you can easily access and run scripts from. *Figure 8.22* shows an example of an inline script created using vim in and run from the Crosh command line:

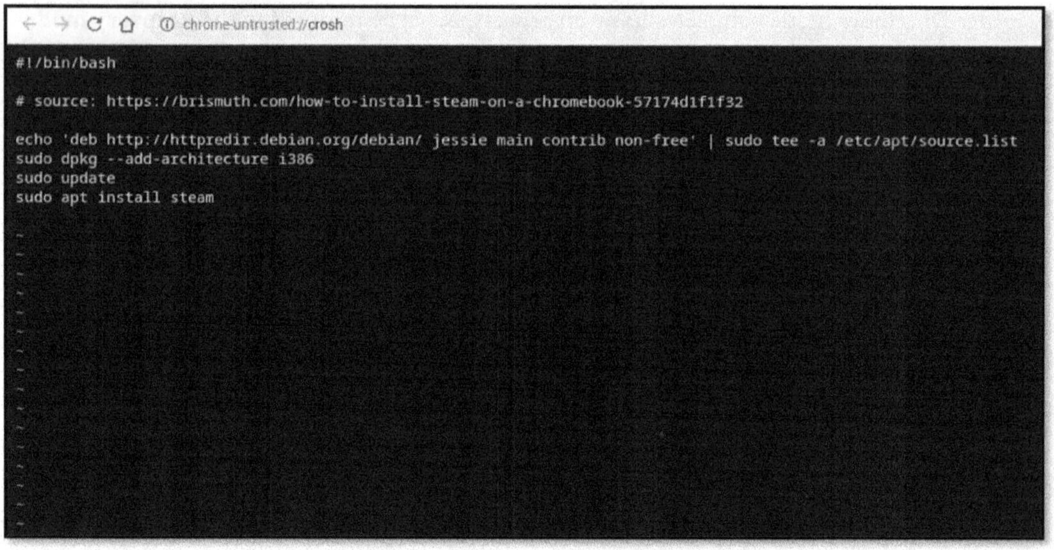

```
#!/bin/bash

# source: https://brismuth.com/how-to-install-steam-on-a-chromebook-57174d1f1f32

echo 'deb http://httpredir.debian.org/debian/ jessie main contrib non-free' | sudo tee -a /etc/apt/source.list
sudo dpkg --add-architecture i386
sudo update
sudo apt install steam
```

Figure 8.22 – Linux script on the Crosh command line

From this point forward, you can begin using the same technique that Linux system administrators use to script administrative tasks. For more information on Linux Shell scripting, check out *Learning Linux Shell Scripting* by Ganesh Sanjiv Naik from Packt Publishing: https://www.packtpub.com/product/learning-linux-shell-scripting-second-edition/9781788993197.

Summary

In this chapter, we took a deep dive into the Crosh command line and learned how to access ChromeOS's unique, browser-based CLI. You got the chance to experiment with a host of system administration commands. You even learned how to enter developer mode, making bash scripting possible. Developing each of these skills is essential as you grow your ability to administer ChromeOS systems at the enterprise level.

In the next chapter, we will move beyond our local ChromeOS device and into the cloud as we explore the **Google Workspace Admin console**.

Google Workspace Admin Console

At this point in your journey to becoming a full-fledged ChromeOS systems administrator, it should be pretty clear how impressive the OS and its native hardware can be. Additionally, with the popularity of Chromebooks on the rise, more and more organizations are beginning to adopt them as their computers of choice. However, even though ChromeOS has powerful local administration tools, it still needs help when introduced into enterprise environments where the centralized management of devices is essential. This is where Google Workspace comes in to elevate your administrative possibilities.

In this chapter, you'll learn how Google Workspace and its Admin console help ChromeOS meet the needs of an enterprise. We'll cover the following topics:

- An overview of Google Workspace
- The major features of the Admin console
- Migrating to Google Workspace

Technical requirements

In order to follow along with the activities outlined in this chapter, you'll need access to the following:

- A device with ChromeOS or ChromeOS Flex installed
- A Google Workspace subscription for a business, school, or nonprofit, or a free trial

An overview of Google Workspace

As you may recall, you were first introduced to Google Workspace way back in *Chapter 3*, *Exploring Google Apps*. There, you learned that Google Workspace is a collection of cloud-based communication, collaboration, and business apps. We discussed how gaining access to one of several types of Google Workspace editions opens the door to accessing advanced application features and resources. You also learned that as a cloud service, Google Workspace uses a *pay-as-you-go* subscription model that if not correctly managed, could result in you spending more than you need on your app access. In short, you learned quite a bit about Google Workspace. However, there is another side to Google Workspace that we are yet to explore: *its role as a hub for the centralized administration of your ChromeOS devices, users, data, and applications*.

The introduction of Google Workspace into your organization's computing environment will provide several benefits when compared to performing management using the local administrative tools found in ChromeOS. These benefits include the following:

- Centralized control over Google Workspace apps such as **Gmail**, **Drive**, **Calendar**, and **Meet**
- Enhanced security features such as client-side encryption
- Larger amounts of storage
- The ability to create **shared drives** in Google Drive
- The easy implementation of IT governance policies
- The ability to generate user and system activity reports

In order to access the feature enhancements listed here, and much more, the first step is to set up a Google Workspace account subscription. This differs from the free Google Account that you've used throughout your exploration of ChromeOS so far. Again, as was eluded to in *Chapter 3*, *Exploring Google Apps*, by opting for a premium subscription to Google Workspace, you'll unlock many of the features held in reserve for paying customers. Let's get started:

1. You need to begin your account setup process by determining what edition of Google Workspace will work best for your organization or business first. Consider things such as the number of users you have, the type of organization you are (for example, a school, a midsized corporation, or a nonprofit), what apps your team will need, how much storage you require, and so on. The subscription editions, which are described in detail earlier in the book, are **Business Starter**, **Business Standard**, **Business Plus**, and **Enterprise**.

2. Once your choice has been made, navigate to Google Workspace's product page at `https://workspace.google.com/` and click on the **Get Started** button, as shown in *Figure 9.1*.

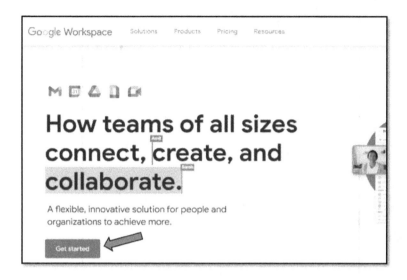

Figure 9.1 – The Get started button on the Google Workspace product page

3. On the **Let's get started** screen, shown in *Figure 9.2*, you'll be asked to provide the following information:

-   **Business name**
-   **Number of employees in your organization, including you:** You will be given several options to choose from, ranging from **Just you** to **300+**

- **Region**: The country your business is based in

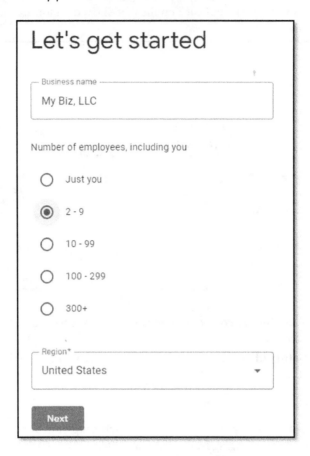

Figure 9.2 – The Let's get started screen of Google Workspace

Provide all the information needed in order to continue with the next steps.

4. After providing all the information, click on the **Next** button, which will take you to the **What's your contact info?** screen. Here, you'll key in your **First name**, **Last name**, and **Current email address** details. This information will be used to establish your Google Workspace Admin account, so be sure to provide a secure email address to link to this account. *Figure 9.3* provides an example of this web form you'll complete:

What's your contact info?

You'll be the Google Workspace account admin since you're creating the account. ⑦

First name

Last name

Current email address

Next

Figure 9.3 – What's your contact info?

5.   Next, you'll be asked whether your business has a domain. This is important since the domain name will be used by Google Workspace to create a customized email address for your organization. As shown in the following image, you have two options at this point – you can click on the **Yes, I have one I can use** button, or the **No, I need one** button.

Does your business have a domain?

You'll need a domain, like *example.com*, to set up email and a Google Workspace account for your business. ⑦

Yes, I have one I can use   No, I need one

Figure 9.4 – Does your business have a domain?

If you already have a website or have paid for a customized email in the past, you already have a lease for a domain. If that is the case, you'd choose the **Yes, I have one I can use** option, and then on the next screen, you'd enter the domain name. If you don't have a domain name, choose the **No, I need one** option. This will take you to the **Let's find a domain name for your business** screen. Here, you can search for your business's name and the search tool will then attempt to create a matching (or similar) domain name. *Figure 9.5* illustrates this tool at work.

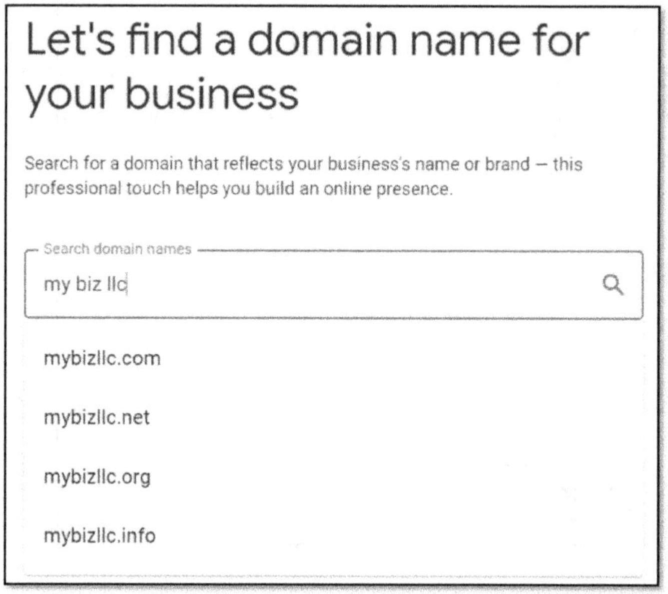

Figure 9.5 – Let's find a domain name for your business

Once you've typed it in, click on the magnifying glass icon to begin the official search for the domain name. The results of this search will indicate whether the domain name is unavailable (that is, already in use by someone else) or available for you to purchase. Additionally, if the domain name is available, the price to lease it for 1 year will also be listed. These prices will vary based on who's providing the domain name lease.

6. Assuming you move forward with purchasing a domain name, next, you'll be asked to enter your business information. This information will be used to register your domain name. *Figure 9.6* illustrates the information that you'll need to provide to move forward with this step.

Figure 9.6 – Enter your business information

7. After submitting your company info, you'll be taken to the **Educate your users** screen. Here, you're given the option to have information about Google Workspace apps automatically sent to your Workspace domain's users after their accounts are created. As shown in the following figure, you can click the **Ok** button if you'd like to opt into this offer or the **No thanks** button to skip it.

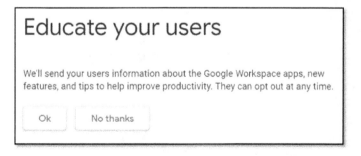

Figure 9.7 – Educate your users

8. Next, on the **How you'll sign in** screen, illustrated in *Figure 9.8*, you'll create a username and password for your Google Workspace account. This username will also be combined with the domain name that you identified or purchased earlier to create your business email address.

Figure 9.8 – How you'll sign in

After clicking on the **Agree and continue** button, you'll be taken to a screen where you can view the payment plan information that you selected. You can also opt into Google's free 14-day trial of Google Workspace or apply promotional discount codes on this screen, as shown in *Figure 9.9*.

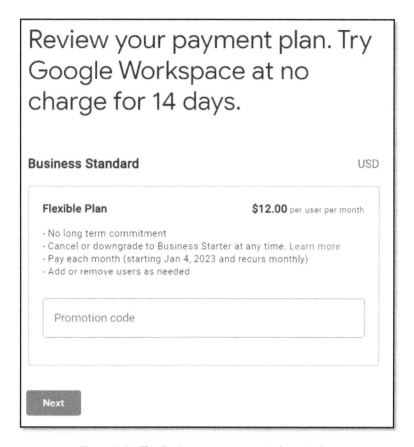

Figure 9.9 – The Review your payment plan window

9. Finally, you'll be taken to a summary screen, which will provide you with all of the details of your order. This is also where you will be prompted to provide payment information for your new account and domain name. You can make changes at this step if you've put in any information incorrectly. Lastly, you can select an option here to make your domain registration information private. This means that the name, address, telephone, and email information you provided during the domain registration process will be hidden from **WHOIS database** searches. By default, this information is viewable by the general public. *Figure 9.10* shows part of the **Review and check out** summary screen. Once you've confirmed that all of your information is correct, click on the **Agree and continue** button to finalize your order.

Figure 9.10 – Review and check out

Now that your Google Workspace account is set up and ready to go, let's examine all of the tools that the **Admin console** has to offer.

The major features of the Admin console

When managing computer devices, there are two widely used approaches. The first approach is the **peer-to-peer (P2P)** model. In this model, the control of computer systems is decentralized. Basically, there is no single system calling the shots. Each device is its own boss, capable of allowing or denying permissions to the resources residing on or attached to the device. You'll see this management approach used in many **Small Office/Home Office (SOHO)** environments due to its easy setup and management.

However, as you start to think more about computing in enterprise environments that can contain large numbers of devices that need to share data, P2P ceases to be a viable option. Instead, you'll commonly find that the **client-server model** is implemented. With this model, you have centralized management via a hierarchical structure. Running the show, you have your servers, which not only provide a centralized place to locate and share resources but also tools for centrally managing access to those resources. *Figure 9.11* gives an example of a common P2P network setup compared to the client-server architecture:

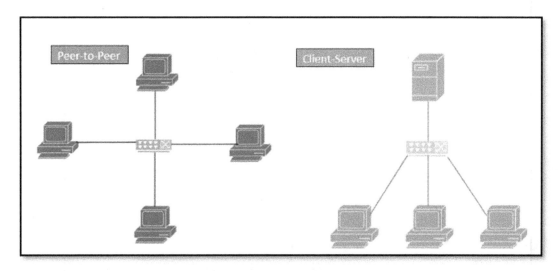

Figure 9.11 – P2P and client-server network examples

You may be thinking, nice networking lesson but what does this have to do with the Admin console? To understand the connection, think about the nature of the cloud. The cloud isn't some mystical, alien technology. It's a mixture of the client-server architecture and virtualization being hosted on the servers of certain companies and leased to you and millions of others for a fee. In short, the cloud is an extension of the client-server model and as such, we can take advantage of its inherent ability to be centrally managed. The Google Workspace Admin console is just Google's tool of choice for allowing administrators to manage their cloud resources, hosted on **Google Cloud Platform** (**GCP**), and on-premises resources (for example, ChromeOS devices) from a centralized management portal.

Now, with that background out of the way, let's learn about Google Workspace's main administrative tool: the **Admin console**.

Accessing the Admin console

Before we can begin to explore the management capabilities of the Google Workspace Admin console, you have to learn how to access it first! Follow these steps to get logged into the Admin console:

1. Use any web browser to navigate to https://admin.google.com.

2. Once you're there, you'll be prompted to log in with a Google account. Use an account login that is connected to a Google Workspace subscription plan.

3. Once your Google account's username and password are confirmed (as well as **Two-Factor Authentication (2FA)** if you have it enabled), you'll have access to the Admin console as seen in *Figure 9.12*.

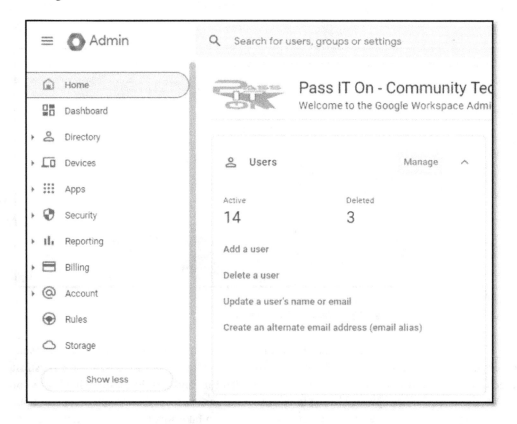

Figure 9.12 – Google Workspace Admin console

The Admin console's home page provides a dashboard-style view of many of its tools. It also contains a left-hand menu that provides a complete list of the tools and features grouped by category. The left-hand menu items are as follows:

*   **Home** – This screen displays the home page screen and returns you to it if you navigate to another screen. The home page contains **Kanban**-style cards that act as a shortcut to the console's frequently used management tools.

*   **Dashboard** – This screen displays the specific status notifications of users and alerts for things such as missing account information and configurations.

- **Directory** – This menu option expands to provide tools to manage users, groups, **Organizational Units (OUs)**, and other domain objects. These options will be explored in depth in *Chapter 10, Centralized Administration of OUs, Users, Groups, and Devices*. Additionally, the menu contains a settings page that controls sharing, profile editing, and OU visibility permissions for the Workspace domain.

- **Devices** – This menu gives an overview of the devices that are connected to your Google Workspace domain. It categorizes them by type; mobile devices, endpoints, Chrome devices, managed browsers, Google Meet hardware, and Jamboard devices. You can also have granular level control over the device with ChromeOS installed on a Chrome web browser. These options will be explored in depth in *Chapter 10, Centralized Administration of OUs, Users, Groups, and Devices*.

- **Apps** – This menu option provides tools that allow you to centrally manage Google Workspace's core services, additional services, web or mobile apps, and Marketplace apps for the domain.

- **Security** – This menu provides tools used to harden authentication, customize alerts, and implement access controls.

- **Reporting** – The menu's main screen provides a dashboard view of many important metrics related to the usage of the Google Workspace subscription. Additionally, the tools grouped under this menu provided canned usage reports, audit logs, and Google Workspace monthly uptime metrics.

- **Billing** – This menu provides information regarding your current subscription and payment account. This is also where you can go to modify your subscription level or purchase add-on features (for example, Google Voice).

- **Account** – This menu allows you to adjust your individual account settings, view or create Admin roles, manage domains, and initiate data migrations into and out of Google Workspace applications.

- **Rules** – This menu allows you to create and manage audit and alert rules for the domain.

- **Storage** – This screen allows you to view the storage utilization of each of the users in your domain and which apps are being used to house the stored data (for example, Drive, Gmail, or Photos). You are also provided with management tools for adjusting the domain's storage and shared drive settings.

Now that you have an overview of what the Admin console has to offer, let's take a deep dive into some of the menu options previously outlined. The **Home** menu doesn't require any additional explanation since it is essentially just a landing page. We can also move past the **Dashboard** menu since it is tailored for each individual user and will vary based on user preference. Additionally, we'll skip the **Directory** and **Devices** menus for now since we will have a more in-depth discussion of these menus in *Chapter 10, Centralized Administration of OUs, Users, Groups, and Devices*. We'll focus our attention on the **Account**, **Billing**, **Reporting**, **Apps**, **Security**, **Storage**, and **Rules** menus.

Account

The **Account** section contains the following submenu structure:

- **Account settings**: The **Account settings** menu, illustrated in *Figure 9.13*, provides a detailed summary of all of your user profile settings, personalization settings (for example, adding a company logo), feature and product preferences (for example, how new features should be rolled out to your account), legal compliances, and custom URLs for Workspace services.

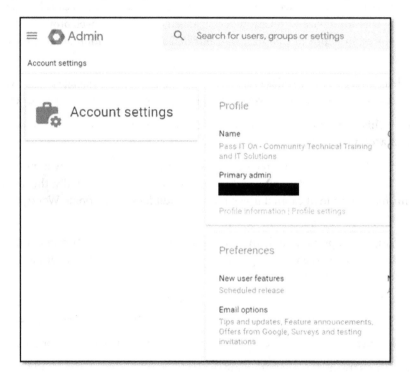

Figure 9.13 – Account settings screen

- **Admin roles**: The **Admin roles** menu provides a listing of the pre-made administrative roles provided by Google Workspace. These roles greatly simplify user administration. They'll be discussed more in *Chapter 10, Centralized Administration of OUs, Users, Groups, and Devices*, when we look at user account creation and management. However, for now, take note of the fact that this menu option allows you to choose from this predefined list or create your own custom roles to meet the specific needs of your organization. *Figure 9.14* shows this menu's available options.

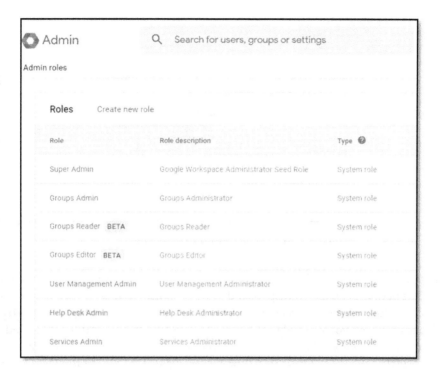

Figure 9.14 – Admin roles menu

- **Domains:** The **Domains** menu has a few submenus of its own: **Overview**, **Manage domains**, and **Allowlisted domains**. On the **Overview** screen, you're able to see two key pieces of information; the number of **managed domains or domain aliases** and the **allowlisted domains**. Managed domains are domains that you've leased to connect to your Google Workspace account for doing things such as creating email addresses and website URLs. The allowlisted domains are trusted domains that belong to a third party but have been granted access to interact with your domain. Note that once you add a domain to the allowlisted domain list, your users will only be able to share documents externally with that trusted domain, so use this feature wisely. The additional menu options under **Domains** give you the ability to manage these items. *Figure 9.15* provides an example of the **Domains → Overview** screen.

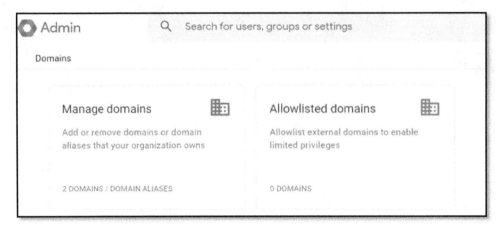

Figure 9.15 – Domains overview screen

- **Data Migration**: Finally, the **Data Migration** screen provides you with the ability to import or export data between Google Workspace and other platforms such as Microsoft Exchange (the email server). The link provide on this screen redirects you to the **Configure Data Migration Service** page, where you can fully configure the details of the migration job, including the migration source, items to be migrated, connection method, and administrative credentials to authorize the migration. *Figure 9.16* shows an example of the **Configure Data Migrations Service** tool.

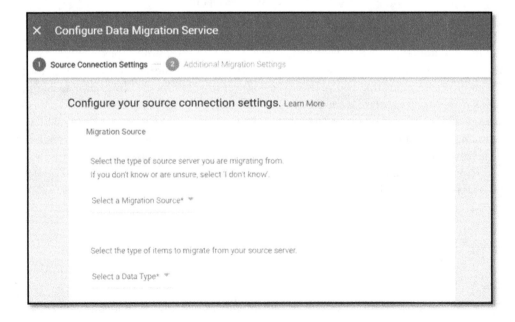

Figure 9.16 – The Configure Data Migration Service screen

Next up, let's look at the **Billing** menu.

Billing

The **Billing** section contains the following submenu structure:

- **Subscriptions**: The **Subscriptions** menu provides a centralized location for all of the Google Workspace subscription information associated with your Google account login. Here, you can see your subscription type, its status (for example, active or inactive), what user accounts are licensed under the subscription, the payment plan, and payment invoices and statuses (for example, in progress, declined, and so on). *Figure 9.17* shows an example of what you'll see when you navigate to this page.

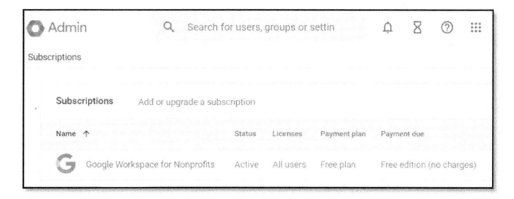

Figure 9.17 – The Subscriptions screen

Additionally, by clicking on the name of your subscription, you can drill down to see how long the subscription has been active or select the option to cancel your subscription.

- **Get more services**: By using the **Add or upgrade a subscription** link found near the top of the **Subscriptions** screen, you are linked to the **Get More Services** page as seen in *Figure 9.18*. There, you can choose to switch to a different subscription plan or even add additional services to your existing plan.

Figure 9.18 – Get More Services

- **Payment accounts**: The **Payment accounts** page is where you manage the payment methods used to pay your subscription fees. The accepted payment methods are via direct debit from a bank account, credit card (Visa, American Express, or MasterCard), or manually prepaid. Businesses also have the option of making invoices, which allows them to pay by check or bank wire transfer.

Next, let's look at reporting in Google Workspace.

Reporting

The **Reporting** section contains the following submenu structure:

- **Highlights**
- **Reports**:
  - **Apps Reports**:
    - **Accounts**
    - **Gmail**
    - **App Script**

- Classroom
- Drive
- Google Meet
- Google Chat
- Voice
- Aggregate reports
- Communities report

- Cost Reports
- User Reports:

    - Accounts
    - Apps usage
    - Security

- Devices:

    - Mobile

- **Audit and investigation:**

    - Admin log events
    - Calendar log events
    - Chat log events
    - Chrome log events
    - Classroom log events
    - Currents log events
    - Data Studio log events
    - Device log events
    - Directory Sync log events
    - Drive log events
    - Graduation log events
    - Groups Enterprise log events
    - Groups log events

- Keep log events
- LDAP log events
- Meet log events
- OAuth log events
- Rule log events
- SAML log events
- Takeout log events
- User log events

- **Manage Reporting Rules**
- **Email Log Search**
- **Google Workspace Apps Monthly Uptime**

As you can see from this list, **Reporting** provides a ton of important data for systems administrators, so let's dig in!

As the name implies, the **Highlights** page provides you with a dashboard full of useful charts and gauges that provide visualizations for your most important data. Each chart also provides a drill-down link that will take you to its associated **app report**, the source of the chart's data. *Figure 9.19* provides an example of what you'll see when you access **Highlights**.

Figure 9.19 – The Highlights screen

The **Reports** menu has several submenus for us to explore, the first being **Apps Reports**. As previously mentioned, this collection of report visualizations acts as the data source for the **Highlights** page. However, each app's reporting screen provides an extremely granular and highly customizable group of reports. Each of these reports can also be exported to Google Sheets or a **Comma-Separated Values (CSV)** file for further data storage and analysis.

Each of your default Google Apps (for example, Gmail, Drive, Docs, Sheets, or Currents) has a collection of reports. Additionally, there is the **Aggregate reports** option. The aggregate reports provide a domain-level trend analysis of app activity that spans all of the Google Workspace app offerings.

The **Cost Reports** page provides a summary of your subscription utilization to help you monitor your service utilization and the associated cost. The **User Reports** menu provides three submenus for viewing and managing user accounts, app usage, and security. *Figure 9.20* shows us the **User Reports → Accounts** page. These reports will be discussed in depth in *Chapter 10, Centralized Administration of OUs, Users, Groups, and Devices*, as we explore the centralized creation and management of user accounts in a domain environment.

Figure 9.20 – User Reports → Accounts screen

The **Devices** menu under the **Reporting** section provides two submenus – **Chrome** and **Mobile**. The Chrome page is used to report on ChromeOS devices that have been enrolled for centralized management (i.e., the client-server model) via the Admin console. The **Mobile** menu tracks the Android and Apple iOS mobile devices that are connected to the Google Workspace domain. *Figure 9.21* shows an example of the visualizations presented on the **Reporting → Devices → Mobile** screen.

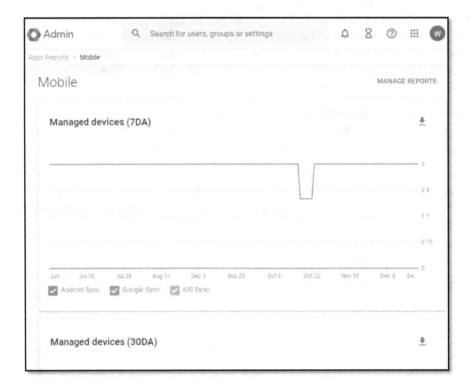

Figure 9.21 – The Mobile report screen

> **Pro tip**
>
> If centralized management of ChromeOS computer devices is a requirement for your computing environment, be sure to choose the right Google Workspace subscription. Currently, only **Enterprise**- and **Education**-level subscriptions allow for device enrollment as a part of their subscription package. However, you can purchase a **Kiosk & Signage Upgrade** or **a standalone upgrade** at an additional cost to enable enrollment capabilities on devices without the required subscription level. Alternatively, you can also upgrade your subscription, although this may be a bit more costly.

The **Audit and investigation** section is yet another submenu, under **Reporting**, that provides several important tools. Here, you'll find all of the log files produced by Google Workspace and the devices associated with your domain. These logs are detailed records of the activities that have occurred on your managed systems, user accounts, and apps. Besides having a log for each of the default Google Apps, you also have device event logs, logs monitoring rules changes, and **Open Authorization** (**OAuth**) logs, tracking application authentication and authorizations. There is even a Graduation log, which allows Google for Education subscribers to track the transfer of student data following their departure from an academic program.

Each log screen gives you the ability to filter the log output based on various parameters (for example, the IP address, event, or date). You can use conditional tests to filter the log output via the **Condition builder** tool (for example, you can create a condition that only shows you results where **Device type** is **Android**). You can also create reporting rules that automatically apply your filters and conditions, as opposed to you being required to create them each time you view the log file. Finally, similar to other Admin console data, the log event files can be exported to Google Sheets or a .csv file for storage and analysis. *Figure 9.22* provides an example of the **Reporting** → **Audit and investigation** → **Admin log events** screen.

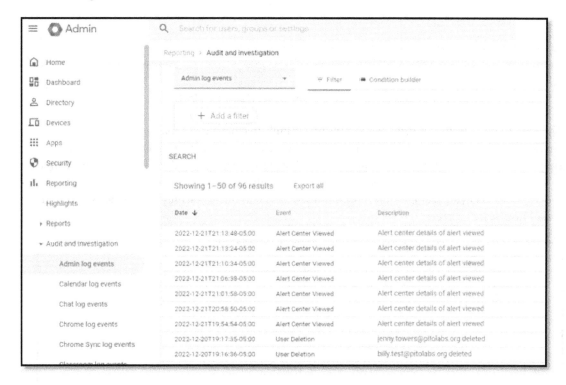

Figure 9.22 – The Admin log events screen

The **Manage Reporting Rules** submenu, seen in *Figure 9.23*, provides a centralized location to view all of the default and user-created rules, alerts, and audit logs that exist in your domain. This screen also provides links to create and investigate rules. These options are actually just links that take you back to the **Audit and investigation** section we previously discussed.

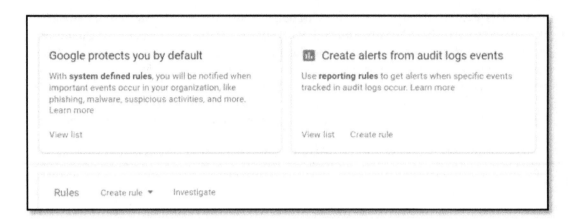

Figure 9.23 – The Manage Reporting Rules screen

The **Email Log Search** page provides admins with a built-in email audit tool. Here, you can search for emails using a combination of filters, including date, sender, sender IP, recipient, recipient IP, email subject, and message ID. The results of your search can then be expanded to allow an administrator to view the details of the message (for example, a list of the email's recipients) and the details of the message's delivery (for example, delivered successfully versus unsuccessfully). *Figure 9.24* depicts the **Email Log Search** screen.

Reports > Email Log Search

Date: Since Yesterday   ⌄   GMT-05:00

Sender: Enter full or partial sender email address   Sender IP: Enter full sender IP address

Recipient: Enter full or partial recipient email address   Recipient IP: Enter full recipient IP address

Subject:

Message ID:

SEARCH
It may take a few minutes for a message to be logged and appear here.

0 results

Figure 9.24 – The Email Log Search screen

Google Workspace Apps Monthly Uptime is the last submenu option we have under **Reporting**. This screen is Google's way of informing users that they are meeting the uptime requirements of their **Service-Level Agreements** (**SLAs**). This page displays not only the overall uptime percentage for Google Workspace on a per-month basis but also the uptime status of each of its core applications. *Figure 9.25* shows part of this reporting screen.

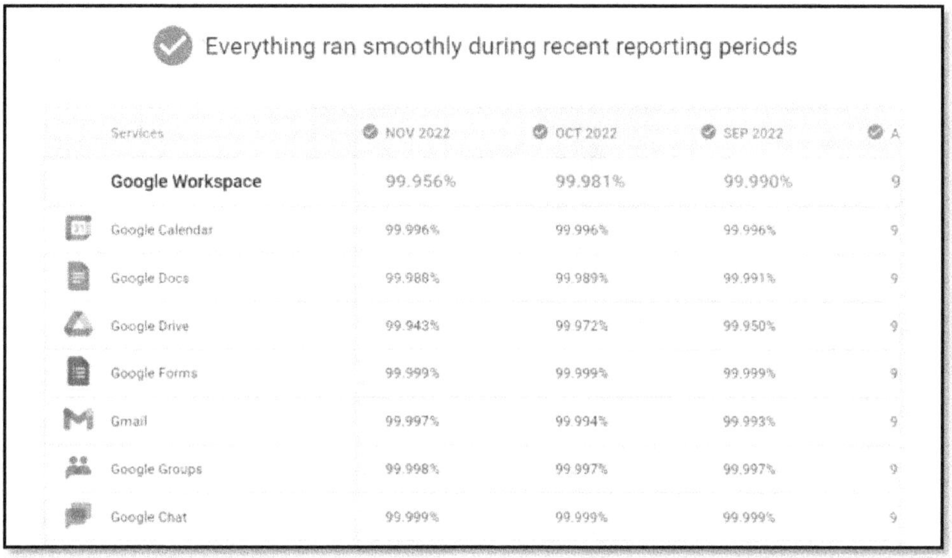

Services	NOV 2022	OCT 2022	SEP 2022	A
Google Workspace	99.956%	99.981%	99.990%	9
Google Calendar	99.996%	99.996%	99.996%	9
Google Docs	99.988%	99.989%	99.991%	9
Google Drive	99.943%	99.972%	99.950%	9
Google Forms	99.999%	99.999%	99.999%	9
Gmail	99.997%	99.994%	99.993%	9
Google Groups	99.998%	99.997%	99.997%	9
Google Chat	99.999%	99.999%	99.999%	9

Everything ran smoothly during recent reporting periods

Figure 9.25 – Google Workspace Apps Monthly Uptime

Apps

The **Apps** section contains the following submenu structure:

- **Overview**: The **Overview** page of the **Apps** menu provides a summary of all of the apps currently deployed in the Google Workspace domain. These apps are grouped into the **Google Workspace**, **Additional Google services**, **Web and mobile apps**, and **Google Workspace Marketplace apps** categories. Each category has a Kanban card, which also doubles as a link to take you to their associated **Apps** submenu. *Figure 9.26* provides an example of what you'll see on this screen.

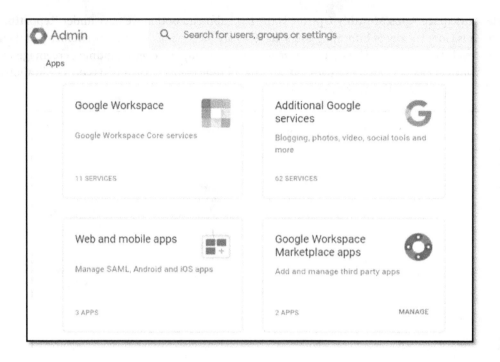

Figure 9.26 – The Apps screen

- **Google Workspace**: This submenu has the following submenus:

    - **Service Status**
    - **Calendar**
    - **Classroom**
    - **Drive and Docs**
    - **Gmail**
    - **Google Chat and classic Hangouts**
    - **Google Meet**
    - **Groups for Business**
    - **Jamboard**
    - **Keep**
    - **Sites**
    - **Tasks**

The **Google Workspace** submenu provides individual settings screens for each of the default Google Workspace apps. These settings menus provide a way to enable or disable an app or control its advanced configurations. These differ from the settings discussed in *Chapter 3, Exploring Google Apps*, because they provide admins with the ability to manage app settings and usage for groups of devices and users, or even the entire domain, from a single screen. The Google Workspace submenu also provides a **Service Status** page that allows you to quickly add and remove app services for all of your domain accounts or specific users and groups. *Figure 9.27* displays a portion of the **Service Status** screen.

Figure 9.27 – The Service Status screen

- **Additional Google Services**: As you learned in *Chapter 3, Exploring Google Apps*, Google offers an entire library of apps on top of the commonly used default app such as **Drive**, **Docs**, and **Gmail**. These apps can also be centrally managed for users, groups, or an entire domain via the **Additional Google services** submenu page. The apps available for management include useful tools such as **Chrome Canvas**, **Blogger**, **Chrome Web Store**, **Chrome Remote Desktop**, **Google Play**, and more. *Figure 9.28* depicts this management page.

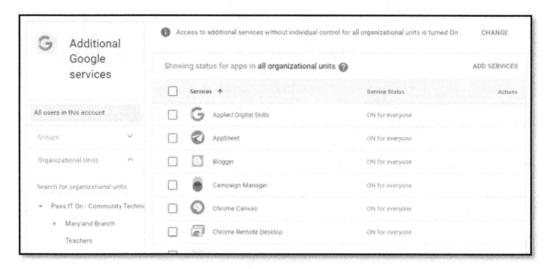

Figure 9.28 – The Additional Google services screen

- **Web and mobile apps**: The **Web and mobile apps** page allows you to view and manage Android and iOS apps that have been approved for users and groups in your domain. Additionally, once an app has been approved for a domain, you can leverage the user access and settings menus, which grant you greater control over how the app is deployed and used. *Figure 9.29* shows an example of an Android mobile app **Settings** page.

Figure 9.29 – The Calculator mobile app settings screen

- **Google Workspace Marketplace apps**: This comes with the following submenus:

  - **App list**

  - **Settings**

The **Google Workspace Marketplace apps** submenu is the last option under the **Apps** menu. It provides users in enterprise environments with the ability to integrate third-party software tools with Google's default apps, making them even more useful in business settings. For example, tools such as **Trello** and **Asana** can integrate with **Gmail** and **Forms** to automate project workflows. The **Google Workspace Marketplace apps** → **Apps list** menu provides you with the ability to view **Domain Installed Apps** and control **Allowlisted Apps** and **Blocked Apps**, which is another way of saying application whitelisting and blacklisting. *Figure 9.30* illustrates what you may see on the **Google Workspace Marketplace apps** → **Apps list** screen.

Figure 9.30 – Apps list screen

The **Google Workspace Marketplace apps** → **Settings** menu is used to allow or block a user's ability to install and run apps from Google Workspace Marketplace. *Figure 9.31* depicts the settings options.

Figure 9.31 – The Google Workspace Marketplace → Settings menu

That's it for the **Apps** menu – now let's explore our security options.

Security

The **Security** section contains the following submenu structure:

- **Overview**: The **Overview** page under the **Security** submenu acts as a portal page to link you to other areas of the Admin console. These links make it easier to access management tools that have an impact on the overall security of your Google Workspace domain. For example, the portal contains a link to the **Rules** page, which allows you to view, create, and manage the protection, alert, and audit rules for your domain. This is the same tool screen previously discussed in the **Reporting** submenu, minus the reporting-specific filtering.

- **Alert center**: The **Alert center** page provides tools that allow you to manage system alerts and email notifications. Here, administrators have the ability to automate monitoring resources, which can help to streamline the investigation of potential security incidents.

- **Authentication**: The **Authentication** submenu has several management pages that control various aspects and methods of user authentication:

  - **2-step verification**: The **2-step verification** screen allows you to enable or disable and configure 2FA at the user, group, or domain level.

  - **Account Recovery**: The **Account Recovery** page lets you configure who can recover a **superadmin** or **user** account password in the event that it is forgotten. Activating self-recovery enables the **Forgot password** option on domain users' login screens. *Figure 9.32* provides an example of the **Account Recovery** page.

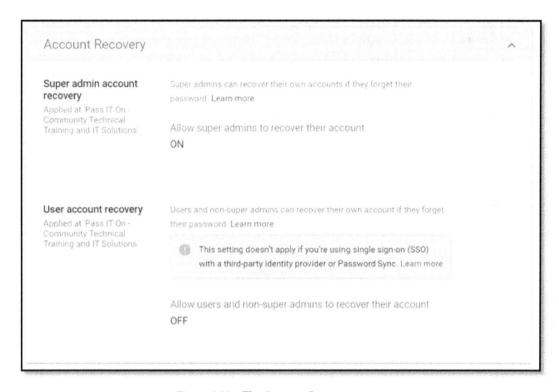

Figure 9.32 – The Account Recovery screen

- **Advanced Protection Program**: The **Advanced Protection Program** menu under the **Authentication** menu provides an extra layer of defense for a user account on your Google Workspace domain that could be a considered high-value target. This includes accounts that have administrative-level privileges on your domain or account belonging to users who frequently handle sensitive data (for example, business executives, teachers, journalists, or medical professionals). This page allows the feature to be enabled or disabled and controls how single-use security codes are managed.

- **Login challenges**: The **Login challenges** page allows you to leverage **Single Sign-On (SSO)** and enable or disable the login challenge feature. This feature's purpose is to provide additional verification of a user's identity in the event of a suspicious login attempt. Challenge questions can include entering an employee ID number, responding to a verification message on a mobile device, or entering a recovery email address. *Figure 9.33* depicts this configuration page.

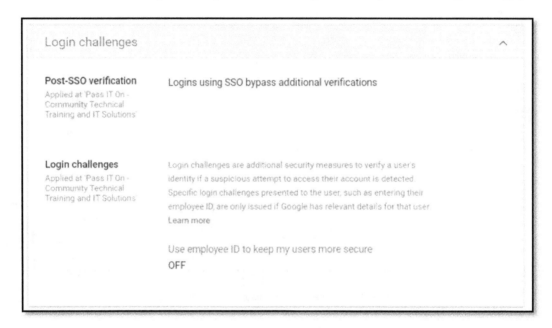

Figure 9.33 – The Login challenges screen

- **Password Management**: The **Password Management** page allows you to implement password rules such as complexity, minimum or maximum length, password reuse limits, and password expiration. *Figure 9.34* depicts these management options.

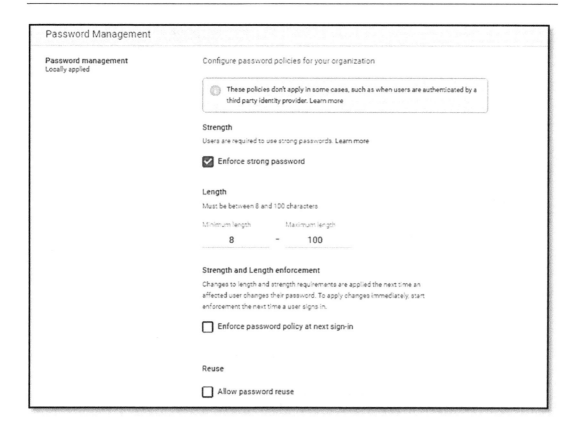

Figure 9.34 – The Password Management screen

- **SSO with SAML applications** and **SSO with third party IdP**: The **SSO with SAML applications** and **SSO with third party IdP** submenu pages allow you to set up SSO cloud applications that support **Security Assertion Markup Language (SAML) 2.0** or **OpenID Connect (OIDC)**. When these configurations have been supplied with the proper settings, they allow you to use your Google Workspace credentials to log into third-party, enterprise cloud applications such as **Tableau**, **SurveyMonkey**, and **Smartsheets**, to name a few.

- **Access and data control**: The **Access and data control** submenu of **Security** also has several configuration pages you should be aware of:

  - **API controls**: The **API controls** page is used to enable or restrict access to Google Workspace's APIs for customer-owned and third-party applications and services accounts. This helps to ensure that only trusted users are able to access the application data produced or residing in your domain's Workspace apps.

- **Client-side encryption**: When enabled, **Client-side encryption** allows **Enterprise-** and **Education**-level Google Workspace users to implement their own encryption key to enhance Google's default encryption. This extra layer of protection encrypts data at the browser level before it is stored on Google Drive servers, meaning that not even Google can access your data or the keys used to encrypt or decrypt it.

- **Data classification**: **Data classification** allows you to create labels that indicate the sensitivity level of files stored on Drive. These labels, listed from lowest to highest sensitivity, are **Public**, **Internal**, **Confidential**, and **Restricted**. Assigning labels allows simplified implementation and the enforcement of data protection policies.

- **Google Cloud session control**: The **Google Cloud session control and SDK session control** page allows you to configure how long a user is able to remain in a Google Cloud console or **Software Development Kit (SDK)** before they're required to sign in again (reauthenticate). The configuration page also allows you to set the method of reauthentication to require a password or security key. *Figure 9.35* depicts this configuration screen.

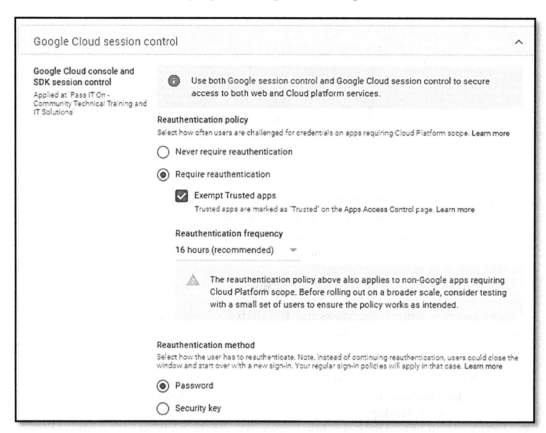

Figure 9.35 – The Google Cloud session control screen

- **Less secure apps**: Lastly, the **Less secure apps** screen provides the option to allow or disable your domain user's ability to access apps with less secure sign-in methods, as they could potentially lead to your account being compromised. Note that if 2FA has been enabled on your system, this feature is enabled by default. Some examples of less secure apps are older versions of iOS, Mac OS X, and older versions of Microsoft Outlook.

Next, let's look at how storage is managed in the Admin console.

Storage

The **Storage** section contains no submenus. Instead, it provides a single dashboard experience that allows you to view and navigate to various storage settings. *Figure 9.36* shows us the **Storage** dashboard.

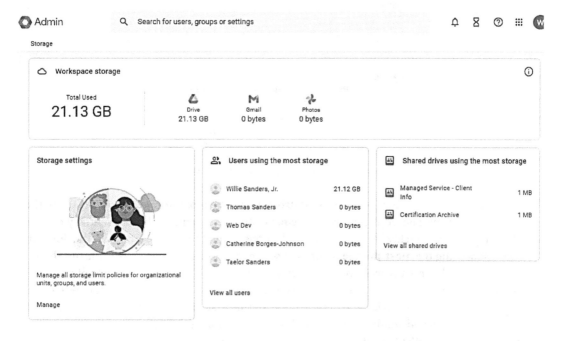

Figure 9.36 – The Storage screen

As you can see, the **Storage** page is separated into four main areas:

- **Workspace storage**: As mentioned previously in this chapter, the **Workspace storage** area provides the overall storage utilization throughout the Workspace domain. It also breaks down this utilization by the apps that actually consume storage in Google Workspace; **Drive**, **Gmail**, and **Photos**.

- **Storage settings**: The **Storage settings** configuration screen allows you to enable or disable storage limits on a per-user or per-OU basis. If limits are enabled, you also are able to control the amount of storage that is allotted. *Figure 9.37* shows an example of the **Storage settings** screen.

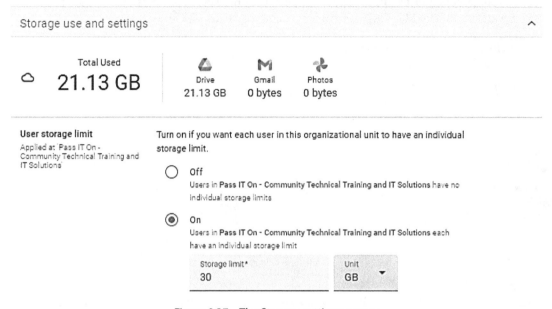

Figure 9.37 – The Storage settings screen

- **Users using the most storage** and **Shared drives using the most storage**: The **Users using the most storage** and **Shared drives using the most storage** screens are both essentially reports that help admins quickly identify who may be consuming too much storage on the domain. The **Users using the most storage** report focuses on individual user accounts while the **Shared drives using the most storage** report focuses on the drives accessible to groups of users. It's important to note that each of these reports is monitoring a separate pool of storage. Google provides users with individual storage quotas per user account based on their Workspace subscription plan and another consolidated storage pool for shared drives based on the overall number of users under the subscription (for example, Enterprise customers get 5 TB of shared storage per user subscription).

Rules

The **Rules** section contains no submenus. In fact, it actually links to the same management tool available under the **Reporting → Manage Reporting Rules** submenu. The key difference is that under **Rules**, the filter that limits the tool to displaying only report-related rules has been removed. This enables administrators to explore and manage the full array of security and performance rules available in Google Workspace. *Figure 9.38* provides a brief sample of the pre-made rule options available to admins, as well as the tool links for creating custom rules.

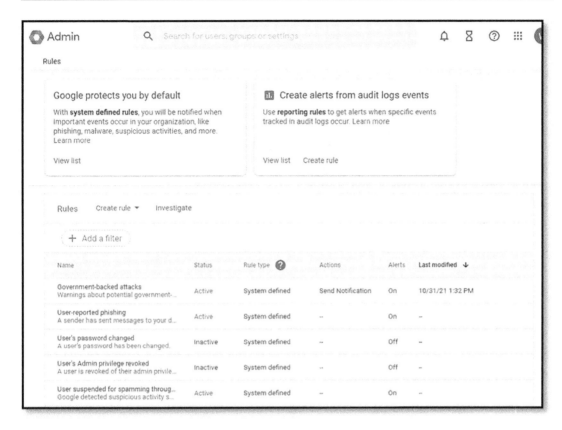

Figure 9.38 – The Rules menu's administrative tool

Now that we've completed our overview of the Google Workspace Admin console, let's determine the best ways to migrate to the platform.

Migrating to Google Workspace

Up to this point, we've been approaching Google Workspace from the perspective of a new subscriber with no previous on-premises or cloud resources or infrastructure at their disposal. A clean slate… a blank canvas. However, a system administrator's reality is seldom that simple and in many cases, you'll need to move data and resources from another platform into Google Workspace. Thankfully, Google has provided several tools and strategies to simplify these migrations.

Most of these migration strategies go beyond the scope of this book. However, Google has provided a host of detailed guides, which provide step-by-step instructions for different migration scenarios. These scenarios include migrating email data, servers, collaboration products, and filesystem data from various platforms to Google Workspace. These guides are located at `https://support.google.com/a/answer/6251069?hl=en`.

Summary

In this chapter, we moved beyond our individual ChromeOS device and began our exploration of Google Workspace's Admin console as an enterprise solution for ChromeOS. You learned what Google Workspace is and how to set up the appropriate subscription to access it. You explored the many features included in the Admin console and learned how to locate and configure several key systems administration features. Finally, you gained some insight into the ability to migrate data from other platforms into Google Workspace, and you were provided with resources, from Google, to help you along the way.

Next, in our final chapter, we will continue to develop a mastery of Google Workspace's administrative capabilities by learning about the tools and techniques for the centralized administration of OUs, users, groups, and devices.

10

Centralized Administration of OUs, Users, Groups, and Devices

As you've learned throughout this book, there are many administrative tools and features that can be utilized to administer an individual ChromeOS device. However, such tools don't provide the scalability needed to handle the administration of these devices when deployed in larger enterprise computing environments. In those instances, you must rely on centralized management tools such as the **Admin Console**, which was discussed in detail in *Chapter 9, Google Workspace Admin Console*.

However, a few areas of the Admin console were excluded from the last chapter's discussion because they require a more in-depth explanation. Specifically, they were the console's **Directory** and **Devices** submenus. These areas of the console contain the tools for administering some of the most important resources in your domain. In fact, the centralized management of these resources is considered to be the hallmark of **enterprise endpoint system administration**.

In this chapter, you'll learn about the following:

- Exploring the **Directory** and **Devices** submenus in the Admin Console

- How to get organized with OUs

- Advanced user account management

- How to manage groups and target audiences

- Advanced device management

- Advanced application management

So, if you're ready to take that final step toward pro-level ChromeOS administration skills, then let's get started!

Technical requirements

In order to follow along with the activities outlined in this chapter, you'll need access to the following:

- A device with ChromeOS or ChromeOS Flex installed

- A Google Workspace Business, School, or Nonprofit subscription or a free trial

Exploring the Directory and Devices submenus in the Admin console

Before we get into the details of managing objects in Google Workspace, let's take a quick look at what the Admin Console's **Directory** and **Devices** submenus have to offer.

Directory

The **Directory** section contains the following submenu structure:

- **Users**: The **Users** section of the **Directory** submenu is your primary tool for managing user account information in Google Workspace. Some of its basic features include the ability to add user accounts individually or in bulk, download lists of user account information, reset account passwords, and delete user accounts. Additional advanced features for managing users will be discussed later in the chapter. *Figure 10.1* shows an example of what you'll see on the **Users** screen:

Figure 10.1 – Users screen

- **Groups**: The **Groups** section of the **Directory** submenu is used to create and inspect groups in Google Workspace. Groups are an important account management tool in systems administration, so their inclusion as a feature of the Admin console is critical. We'll talk more about groups later in this chapter, but for now, take a look at their management screen, presented in *Figure 10.2*.

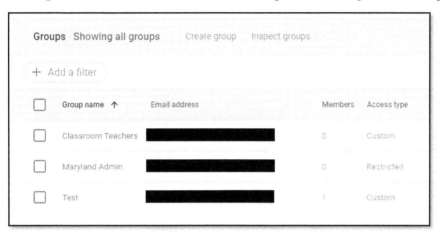

Figure 10.2 – Groups screen

- **Target audiences**: The **Target audiences** section of the **Directory** submenu is a relatively new addition to Google Workspace's administrative toolset as of the writing of this book. Its purpose is to help improve the sharing of content via the **Google Drive**, **Docs**, and **Chat** services. To accomplish this, admins are able to create a special kind of group called a target audience. More on this later in the chapter. *Figure 10.3* illustrates the **Target audiences** management console.

Target audiences | Showing all target audiences		Create a target audience
Name	Members	Description
Pass IT On - Community Technical Training and IT Solutions	14	Default audie
Teachers	1	This group is
Sys Admins	0	System Admi

Figure 10.3 – Target audiences screen

- **Organizational units**: The **Organizational units** section of the **Directory** submenu gives you control over one of the key building blocks of any domain structure, the **Organizational Unit (OU)**. This section of the Admin console allows you to create, manage, and delete OUs. Very soon, we'll explore this process step by step, but for now, take a look at the OU management screen.

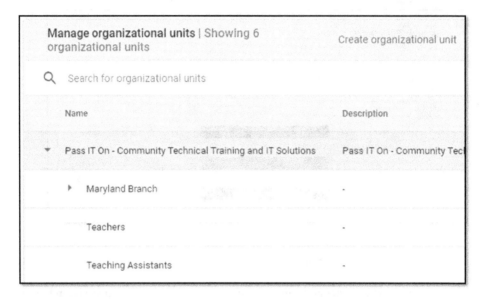

Figure 10. 4 – Organizational units screen

- **Buildings and resources**: The **Buildings and resources** section of the **Directory** submenu is a great illustration of how everything in the Google Workspace directory structure is treated as an object to be managed, including buildings and rooms. The subsections of **Buildings and resources** provide an overview of the currently defined building and room resources managed via your Google Workspace environment:

  - **Overview**
  - **Manage resources**
  - **Room insights**
  - **Room settings**

  The **Manage resources** screen allows you to add building and resource details to the Admin console. **Room insights** allows you to view the availability and usage patterns of your room resources. You can use this screen to book the usage of room resources. Finally, the **Global room settings** screen, seen in *Figure 10.5*, allows you to configure calendar-based room releases, exempt user groups, and automatic room replacement.

Figure 10.5 – Global room settings screen

Calendar-based room release helps to simplify room resource management by automatically releasing a room resource reservation if no one but the meeting organizer accepts the meeting invite. This release allows the room to show up as available on the **Resource insights** screen as well as in **Google Calendar**. The **Exempt user group** option allows you to make exceptions to **Calendar-based room release** by identifying user groups whose reservations will remain on the room resource calendar, regardless of how many people accept or decline the invitation (for example, manager or director groups). The **Automatic room replacement** feature streamlines the process of identifying an alternative room if your room reservation is declined. When this feature is enabled, the system will automatically substitute your rejected room selection for a room resource of comparable size and with comparable equipment based on the details entered on the **Manage resources** screen.

- **Directory settings**: The **Directory settings** section of the **Directory** submenu allows you to manage the global directory service settings for your entire Google Workspace domain. These settings fall into three categories: **Sharing settings**, **Visibility settings**, and **Profile editing**. **Sharing settings** allows you to define whether contacts can be shared and whether data can be shared externally or publicly. **Visibility settings** lets you define what user accounts in your OU different users are able to see. Your options are to allow no users, all users, or a custom set of users to see other members of their OU. **Profile editing** allows you to define whether users are allowed to edit their own user profiles. You also can define what information will be visible in the user profile, as illustrated in *Figure 10.6*.

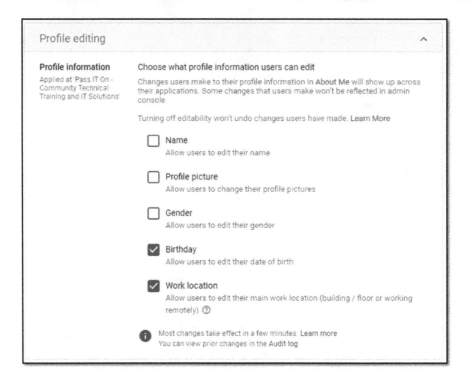

Figure 10.6 – Profile information screen

- **Directory sync**: This section has the following submenus:

  - **Overview**

  - **LDAP directories**

  The **Directory sync** section of the **Directory** submenu, which is in beta testing as of the writing of this book, provides an overview and the ability to sync LDAP directories to objects in Google Workspace. **Lightweight Directory Access Protocol** (**LDAP**) is a set of rules and processes that define how applications can rapidly query and share user information. Unlike technologies such as **Microsoft's Active Directory**, LDAP is vendor-neutral and compatible with practically every directory management technology available. This makes it the perfect solution for sharing and authenticating user information between Google Workspace and other applications and systems. The LDAP directories screen allows you to connect Google Workspace to other LDAP-compatible directory services such as Active Directory and **OpenLDAP**.

Now that you've had an overview of the Admin console's **Directory** submenu, let's look at the **Devices** submenu.

Devices

The **Devices** section contains the following submenu structure:

- **Overview**
- **Chrome:**
  - **Overview**
  - **Guides**
  - **Devices**
  - **Managed browsers**
  - **Settings:**
    - **Users & browsers**
    - **Device**
    - **Managed guest sessions**
  - **App Extensions:**
    - **Overview**
    - **Users & browsers**
    - **Kiosks**
    - **Requests**
  - **Connectors**
  - **Printers:**
    - **Printers**
    - **Print servers**
  - **Reports:**
    - **Overview**
    - **Devices**
    - **Versions**
    - **Apps & extensions usage**
    - **Android app installation**

- Insights
- Printers

- Mobile & endpoints:

  - Devices
  - Company owned inventory
  - Device approvals
  - Settings:

    - Android settings
    - iOS settings
    - Windows settings
    - Universal settings
    - Third-party integrations

  - Reports
  - Manage Reporting Rules
  - Rules

- Networks

As you can see from the breakdown, there's a lot to unpack in the section. So let's get to it!

The main **Overview** section of the **Devices** submenu provides a summary dashboard that displays the major groupings of devices that exist in the Google Workspace environment. A Kanban board is used to represent each device category. The boards display the number of devices currently associated with each category. The Kanban boards also act as links, which can then be used to jump to other areas of the Admin console. *Figure 10.7* illustrates what you might see when you access the **Devices → Overview** screen.

Figure 10.7 – Devices → Overview screen

The **Chrome** section of the **Devices** submenu provides its own **Overview** screen. Here, you can access links to a myriad of granular-level configuration options and help menus. Under the **Chrome →** **Guides** section, you are provided with brief summaries on how to set up a task for ChromeOS and the Chrome browser along with links to each task's respective management page. The **Chrome → Devices** section allows you to see computer devices that have been enrolled for centralized management in Google Workspace. This screen also allows you to initiate **Manual** and **Zero-touch** enrollments, as illustrated in *Figure 10.8*.

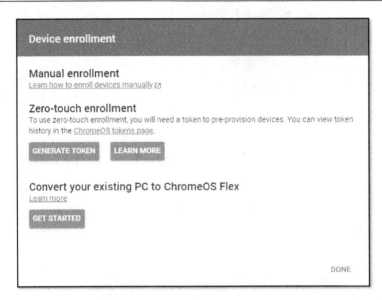

Figure 10.8 – Device enrollment screen

> **Pro tip**
>
> When performing enterprise endpoint management, keeping a proper inventory of your computer assets is extremely important. Therefore, to make your job a bit easier, the **Chrome → Devices** page not only does enrollments, but it also allows you to export your asset list to a .csv file and even view/upgrade their device subscription levels. You're welcome!

The **Chrome → Managed browsers** section of the **Devices** menu gives you access to an administration option unique to Google – the ability to manage the Chrome web browser across multiple OS platforms. From this menu, you can enroll and centrally manage Chrome browser settings, extensions, and policies for not only ChromeOS devices but for any Windows, macOS, iOS, or Android device running the browser.

The **Chrome → Settings** section of the **Devices** menu includes advanced global settings for the authenticated user and managed guests, the browser, and devices. How to configure these options will be discussed later in the chapter.

The **Chrome → Apps & extensions** section of the **Devices** menu provides insight into the **Google Play Store** and **Chrome Web Store** apps that have been installed on computers, in managed browsers, and on devices configured as kiosks in your Google Workspace environment. You can also view and manage user requests for Chrome browser extensions from this screen. The **Users & browsers** tab on this screen is of special interest because it contains the **Additional Settings** link, which can be used to customize the app's user experience for users, groups, or entire OUs. We'll explore these settings in depth later in the chapter.

The **Connectors** section of the **Devices** submenu, when enabled, allows you to connect Chrome to various third-party software tools. Software from companies such as Splunk and CrowdStrike can be integrated to provide reporting and content analysis on your Google Workspace user's browser activity.

Figure 10.9 – The Connectors screen

The **Chrome → Printers** section of the **Devices** submenu provides a centralized management screen for printers and print serves in Google Workspace. Here, you can view these devices or add them to your environment individually or in bulk by uploading a list of printer names and configurations in a .csv file. *Figure 10.10* provides an example of the **Add Printers** configuration screen.

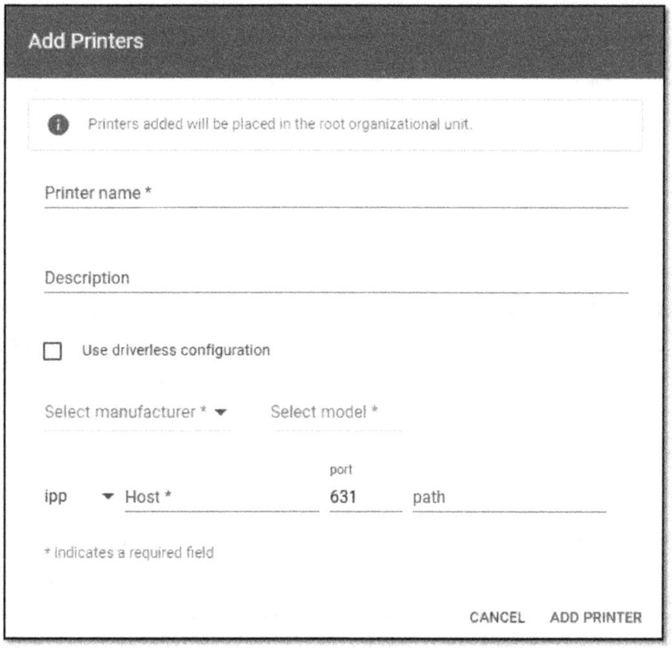

Figure 10.10 – The Add Printers screen

The **Chrome → Reports** section of the **Devices** submenu provides several tabular and graphical reports for various browser, app, and device metrics. The **Overview** screen provides a summary of the available Chrome reports with linked Kanban boards. Each of the reports also provides the admins with the ability to customize and filter them to suit their specific data needs. *Figure 10.11* shows an example of the **Chrome → Reports → Overview** screen.

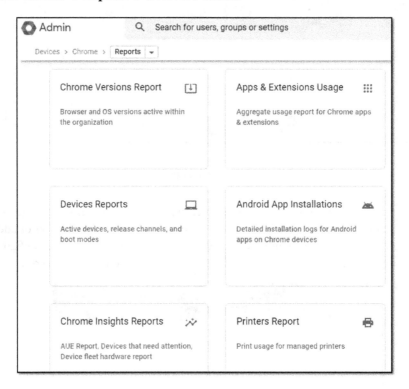

Figure 10.11 – Chrome → Reports → Overview screen

The **Mobile & endpoints → Devices** section of the **Devices** submenu has several screens, which allow admins to track devices that have been used to authenticate to Google Workspace. The **Mobile & endpoints → Company owned inventory** section provides the ability to track both company-owned equipment that is currently assigned to users and still in an organization's possession. The **Mobile & endpoints → Device approvals** section shows the details and statuses of Google Workspace device access approvals.

The **Mobile & endpoints → Settings** section of the **Devices** submenu has several subsections of its own. Each section provides an advanced setting for supported operating systems including Android, iOS, and Windows. In addition, the **Universal settings** screen contains configurations that affect all mobile devices and endpoints that authenticate to the Google Workspace environment. Lastly, the

Third-party integrations screen allows administrators to connect Google Workspace to third-party **Enterprise Mobility Manager (EMM)** software in order to provide centralized management of Android apps via the Admin console as this feature isn't available natively in Google endpoint management.

The **Mobile & endpoints → Reports** section of the **Devices** submenu provides a number of configurable, graphical reports. The reports deliver data on managed mobile devices and devices that have opted into the Google Chrome browser Sync feature. *Figure 10.12* provides one of the reports provided by the **Reports** section. This particular report shows the activity-managed devices that have synchronized with Google Workspace in the last seven days. Note that these reports are the same ones that we viewed when discussing the main **Reports** submenu in *Chapter 9, Google Workspace Admin Console*.

Figure 10.12 – Managed devices (7DA) report from the Devices → Reports screen

The **Mobile & endpoints → Manage Reporting Rules** and **Mobile & endpoints → Rules** sections of the **Devices** submenu link back to the main **Reports** submenu, providing the ability to create and manage reporting rules as well as to investigate alerts. The main difference between these two sections is the fact that the utilities are preconfigured to address reporting and mobile device rules and alerts.

Finally, the **Networks** section of the **Devices** submenu allows administrators to view existing and configure new Wi-Fi, Ethernet, VPN, and cellular network connections. You can also manage network security certificates and help orchestrate their delivery via the creation of **Simple Certificate Enrollment Protocol (SCEP)** profiles. *Figure 10.13* provides an example of this configuration screen.

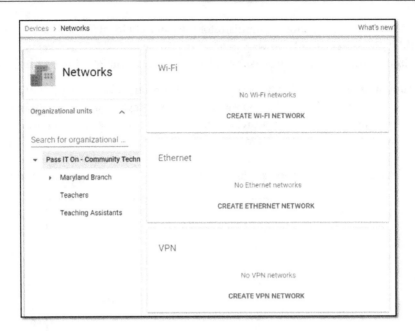

Figure 10.13 – Devices → Networks screen

Now that you've made it through our tour of the **Directory** and **Devices** submenus of the Admin console, let's use your newfound knowledge to implement some advanced configurations! First up, let's start by creating some OUs!

Getting organized with OUs

In most client-server architectures, the domain is the highest-level organizational structure in a directory hierarchy. It usually represents an entire corporate network or at least a major portion of one. However, right under this major structure, we have the **Organizational Unit** (**OU**), which provides another way of grouping objects such as users, groups, computers, and other domain objects into logical groupings. By doing this, an administration can simplify the process of locating devices and user accounts as the domain grows. They can even use the OU structure as a way of segmenting administrative power over the domain by providing some admins access to certain OUs and not others.

Let's use the following steps to create an OU structure in the Admin console:

1.   Open your Chrome browser and navigate to `https://admin.google.com`.

     Note that you will need to log in with a Google account that has admin rights to the Admin console.

2.   Once you're in the Admin console, go to the **Directory** → **Organizational units** section.

3.   When the **Organizational units** screen displays, click the **Create organizational unit** link found near the top of the window, as seen in the following screenshot.

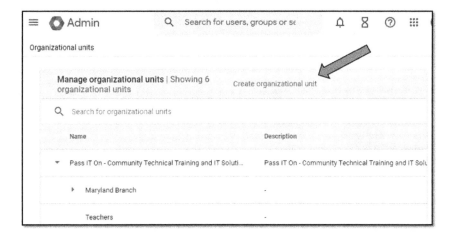

Figure 10.14 – Create organizational unit link

4. Once the link is clicked, the **Create new organizational unit** configuration window will open. Here, you can enter the name of your new OU, provide a brief description of how it will be used, and define which domain or OU will be its parent. That's right, OUs can hold other OUs! *Figure 10.15* provides an example of the **Create new organizational unit** screen. Once you've entered your settings, click the **CREATE** link.

Figure 10.15 – The Create new organizational unit screen

5. Once the OU has been created, it will appear on the **Manage organizational units** screen, as seen in *Figure 10.16*. From there, you can do the following:

- Use the plus sign (+) by the OU's name to add sub-OUs underneath it

- Use the folder/arrow icon to move existing OUs to make them parent or sub-OUs

- Use the kebab menu (three dots) to edit OU details or delete the OU

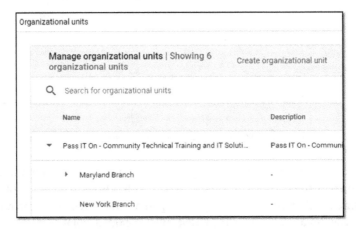

Figure 10.16 – Options for managing OUs

By using these OU management steps, a systems administrator can create an OU structure that mirrors the physical or administrative structure of their organization. Next, we'll look at how user accounts can be incorporated into this hierarchy and some of the advanced configurations an admin can use to configure and manage them.

Exploring advanced user account management

Way back in *Chapter 1, ChromeOS Basics*, you got your first experience of creating Google accounts. A Google account is your primary means of identity management on ChromeOS devices as well as in Google Workspace. However, when you or your organization decides to level up your Google Workspace instance by acquiring paid subscriptions, you unlock the ability to create and centrally manage Google accounts from the Admin console. Earlier in this chapter, you briefly looked at the **Directory → Users** submenu, but now it's time to go all in!

To create a user account, perform the following steps:

1. Open your Chrome browser and navigate to `https://admin.google.com`.

    Note that you will need to log in to the Admin console with a Google account that has admin rights.

2. Once you're in the Admin console, go to the **Directory → Users** section.

3. From this screen, in the **All organizations** section (**1** in *Figure 10.17*), select the organization and OU that you want to create the user account in, then click the **Add new user** (**3**) link. *Figure 10.17* illustrates each of these tasks:

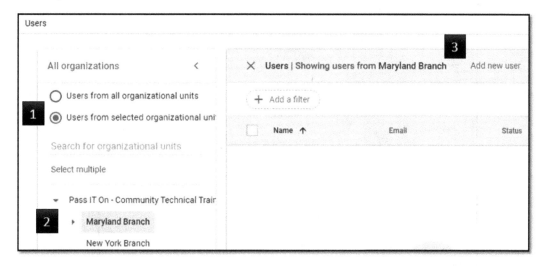

Figure 10.17 – Steps to access the Add new user link

4. Once the **Add new user** link is selected, the **Add new user** configuration screen will display. On this screen, you'll be required to provide the following information for the user:

    - First name

    - Last name

    - Primary email address (on your Google Workspace domain)

    Note that there are also several optional pieces of data that can be entered for the user, including a second email address, phone number, photo, and password (autogenerated or created by the admin).

5. Once all of the required fields of the **User Information** form have been completed, you can click the **ADD NEW USER** button, as seen in *Figure 10.18*, to create the new account.

Figure 10.18 – The User Information form

6. After you click on the **ADD NEW USER** button, you'll be taken to a summary screen that displays the name and password of the account you just created. Here, you can copy or print the password to share with your new user. Alternatively, you can use this page to send sign-in instructions to the user via email (see *Figure 10.19*). Once you've made your decision about what to do at this point, click the **DONE** button to finalize the account creation.

Note that you may need to refresh your web browser or even wait up to 24 hours before the user account becomes visible on your **Users** screen.

Figure 10.19 – The Add New User summary screen

7. Once the user account is created, the user's name will be presented as a hyperlink on the **Users** screen. Click the link to access that account's advanced configurations. The advanced options are split into the following groups:

- **User information**: This is where you get to enter all of the user's detailed information (for example, phone number, date of birth, employee ID number, and so on).

- **Security**: This section provides you with one way to reset a user password and assign encryption keys. You can also enable/disable advanced protection and two-step verification, security features discussed in *Chapter 4, ChromeOS Security*.

- **Groups**: This section lets you see what Google Workspace groups the user belongs to. You can also add them to groups from this section. We'll explore groups later in the chapter.

- **Admin roles and privileges**: This section allows you to assign role-based administrative privileges to users. The roles range from **Super Admin** (an account that has total control of Google Workspace) to **Group's Reader** (a role that simply gives you read-only access to a group's information. There are several other roles with varying levels of administrative power between Super Admin and Group's Reader. Additionally, this screen also provides you with a link to create custom admin roles.

- **Apps**: This section allows you to view the apps that the user has access to and their status.

- **Managed devices**: This section lets you see the managed mobile devices that are assigned to the user.

- **Licenses**: This section lets you see the Google Workspace and app licenses that have been assigned to the user and their estimated monthly cost.

- **Shared drives**: This section identifies any shared drives that the user has access to.

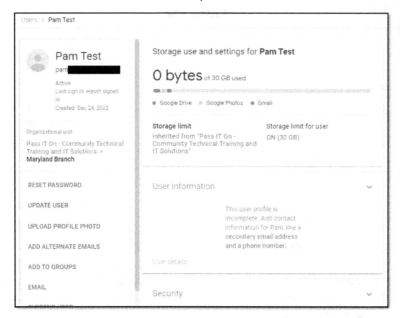

Figure 10.20 – Portion of the Users advanced configuration screen

As you can see, Google provides a large number of options to suit any organization's user management needs. Once data is entered into the user account's fields, it becomes available for use by many of the other tools and features in Google Workspace. So, be as thorough as possible! It will benefit you and your organization in the long run.

Adding or editing multiple user accounts in your Google Workspace

In the event that you need to add or edit multiple user accounts in your Google Workspace environment all at once, you can use the **Bulk update users** option. To do this, perform the following steps:

1. From the **Directory** → **Users** screen in the Admin console, click on the **Bulk update users** link highlighted in *Figure 10.21*.

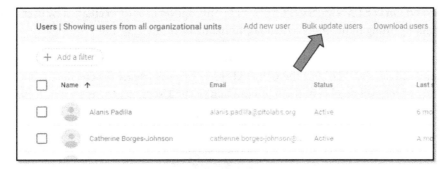

Figure 10.21 – Bulk update users link

2. When the **Bulk update users** screen displays, you'll be provided with the option to either download your existing user account info in a `.csv` file format or download a blank `.csv` template. *Figure 10.22* illustrates each option and the steps that follow in the download process.

Figure 10.22 – Bulk update users screen

The following options are available under **Bulk update users**:

- **DOWNLOAD USER INFO IN CSV FILE**: This option is used to make bulk edits to existing user account entries. The downloaded file will contain the account information for every account in the organization or OU.

- **DOWNLOAD BLANK CSV TEMPLATE:** This option is used to create new user accounts in bulk by filling in the required and optional fields of the CSV file.

3. After the appropriate CSV file has been filled out, use the **ATTACH CSV FILE** button to attach the completed file and then click the **UPLOAD** link. This will trigger the creation of the new user accounts, which will display in the main portion of the **Users** screen, as seen in *Figure 10.23*.

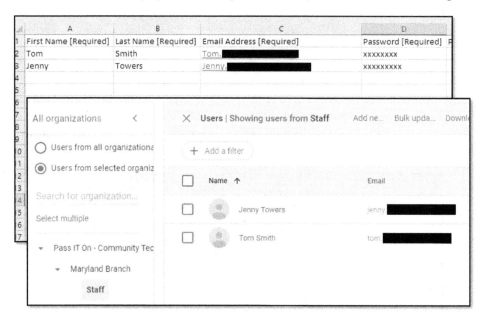

Figure 10.23 – Bulk update users CSV file and its results

Pro tip

If an error occurs during your bulk update users action, it's likely due to incorrect formatting in your CSV file. One easy way to ensure that you provide all of the necessary data in the correct format within the CSV file template is to first download a copy of your existing accounts, created using the GUI tool. By using this document's formatting as a frame of reference, you'll be able to avoid any further format errors.

Managing the custom attributes for user accounts

With your user account now in place, you have the ability to configure the type of information the user profile will hold and present. You also can control whether these settings are managed by individual users moving forward or by an admin such as yourself. To manage the custom attributes for user accounts, perform the following steps:

1. From the **Directory** → **Users** screen in the Admin console, click on the **More options** drop-down link and select the **Manage custom attributes** option.

2. On the **Manage custom attributes** screen, you'll be presented with a list of standard (non-editable) attributes, grouped into categories. You also have access to the **ADD CUSTOM ATTRIBUTE** link, highlighted in *Figure 10.24.*

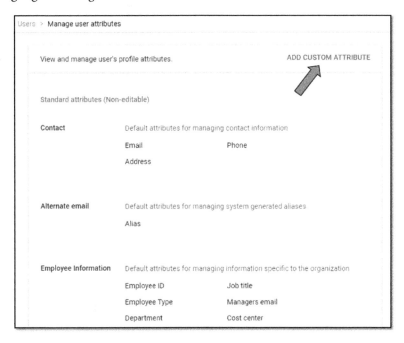

Figure 10.24 – The ADD CUSTOM ATTRIBUTE link

3. The link opens the **Add custom fields** form. Here, you are able to do the following:

- Create a new attribute category

- Add a description of your category

- Enter the name of your new custom field

- Select an info type (for example, **Text**, **Whole Number**, **Yes/No**, **Decimal number**, **Phone number**, **Email**, or **Date**)

- Select a visibility level (**Visible to user/admins** or **Visible to organization**)

- Select the number of values the field will hold (multi-value or **Single value**)

Enter the custom fields information as illustrated in *Figure 10.25* and click the **Add** button to create the new category and fields. These fields will become visible in the user's advanced configurations, under the **User details** section, which was discussed earlier in this chapter.

Figure 10.25 – The Add custom fields screen

Now that we have the user account in place with advanced configurations, let's simplify its management by introducing groups into the workspace.

Understanding groups and target audiences

Grouping objects has been a tried and true method of reducing administrative overhead in organizations for decades. Grouping together users or devices with similar needs (for example, apps, privileges, storage access, and so on) makes it easier to keep track of who can see or do things with or to your resources. Google Workspace provides two versions of groups, one that comes with your free Google account and the other that is exclusively for Google Workspace subscribers. Both versions allow you to collaborate with others but only the subscriber version allows the use of corporate features such as shared group email inboxes, the use of custom domains, and centralized moderation policy administration.

To create a new group, perform the following steps:

1. Navigate to the **Directory → Groups** screen in the Admin console and on the **Groups** screen, click the **Create group** link.

2. When the **Create group** window opens, you'll be presented with the **Group information** screen, as seen in *Figure 10.26*. Enter the details of the new group, including its name, email, description, and owners. You will also define whether the group is for mail distribution, managing security-related settings, or both. After you fill in the required fields, click the **NEXT** button.

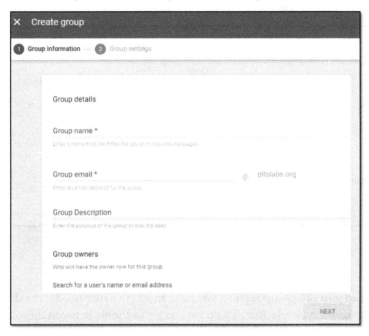

Figure 10.26 – Group information screen

3. Next, on the **Group settings** screen, you'll define the **Access type** for the group. The access type setting determines how group users are able to gain access to groups and what they can do once that access is obtained. Make your selection of one of the options before moving on to the next part of the screen. The options include the following:

-   **Public**: This option allows anyone in your organization to post to or join the group

-   **Team**: This option allows anyone in your organization to post to groups but they have to ask permission to join

-   **Announcement Only**: This option allows only group owners and managers to post to a group but allows anyone in your organization to join

-   **Restricted**: This option allows only group owners, managers, and members to post to the group and requires people in your organization to ask to join

-   **Custom**: This option is automatically set when you alter one of the previously mentioned access types

4. Each access type has a group of associated settings. These settings, shown in *Figure 10.27*, display a matrix of permissions that control who can perform certain actions based on their group membership. These permissions can be modified by checking/unchecking the boxes that align with the group membership level and actions you wish to affect. Make your modifications and move on to the next portion of the screen.

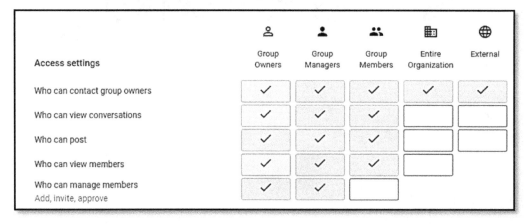

Figure 10.27 – Access settings matrix

5. In the **Who can join the group** section, you have another chance to alter the default attributes of your initial **Access type** selection. This time, you get to choose between the following options:

    *   Anyone in the organization can ask to join a group

    *   Anyone in the organization can join without asking

    *   Joining is done by invite only

    After making your selection, you can move on to the final section of the configuration screen.

6. In the **Allow members outside your organization** section, you are able to toggle on or off a switch that controls whether users from outside of your organization can join your groups. Make your selection and then click the **Create group** button.

7. Once the group is created and the settings have been saved, you'll be taken to the final **Create group** screen. Here, you can add members to your group, review your settings, create another group, or simply click the **DONE** button as illustrated in *Figure 10.28* to close the window.

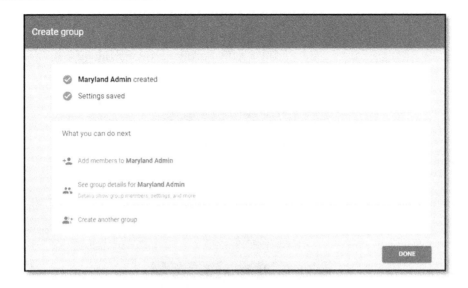

Figure 10.28 – The final Create group screen

Once the group is created, it will appear on your **Groups** screen. From there, you can add members, modify permissions, run inspections to audit group membership, or even delete the group totally.

Another way of sharing resources among a group of users with similar access needs is through the use of **Target audiences**. Target audiences are a specialized type of group that can be leveraged when you need to administer sharing settings for Google Workspace services such as **Drive**, **Docs**, or **Chat**. By using target audiences, you are able to increase the level of data security and privacy of these select Workspace services while simultaneously simplifying the overall process of share settings management. To create target audience groups, perform the following steps:

1. Navigate to the **Directory** → **Target audiences** screen in the Admin console and click on the **Create a target audience** link.

2. When the **Create a target audience** window opens, enter the name in the **Name** field and an optional description of your audience in the **Description** field, as seen in *Figure 10.29*. After the form is filled in, click the **Create** button.

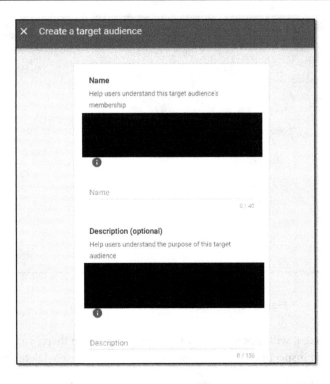

Figure 10.29 – The Create a target audience screen

3. Once the target audience is created, you'll be taken to a summary screen where you have the chance to add members (for example, users or groups) to the target audience and apply the target audience to Google services. You can also create another target audience or simply click the **Done** button to finalize your current target audience creation task.

Now that you've learned how to set up groups and target audiences for administering users in Google Workspace, let's move on to managing devices.

Exploring advanced device management

As an administrator, management of your computing environment extends beyond just users and their groups. You also need to be able to centrally manage the computing devices that connect to your domain. Google Workspace gives you this ability by providing a suite of tools in the **Devices** submenu in the **Admin console**.

To access and manage these configurations, follow these steps:

1. Navigate to the **Chrome → Devices** screen in the **Admin console**, as illustrated in *Figure 10.30*.

Figure 10.30 – The Chrome Devices screen

This screen provides you with the ability to perform the following actions:

- **Enroll**: This option launches the device enrollment tool discussed earlier in the chapter. It allows you to manually add devices to your domain, create zero-touch enrollment tokens, or get information on converting a PC or Mac to ChromeOS Flex.

- **Export**: This option allows you to download a list of enrolled devices to a `.csv` file.

- **Upgrades**: This option allows you to view the Google Workspace licenses assigned to devices in your environment. It also provides shortcuts for getting additional information on license purchases.

Note that to move forward with device management, you'll need to have at least one enrolled device.

2. Once you have an enrolled device, navigate to **Devices → Chrome → Settings → Device**. Here, you'll be given access to an extensive list of configuration options, divided into several groups. Use the following groups to customize various aspects of your managed devices:

- **Enrollment and access**: This group contains settings that control how devices are allowed to enroll, Powerwash usage options, and verified access and mode requirements

- **Sign-in settings**: This group has a collection of settings that control how and when, and from where, a user is allowed to sign in on a managed device

- **Sign-in screen accessibility**: This group's settings allow you to configure tools to assist the blind and visually impaired in using enrolled ChromeOS devices

- **Device update settings**: These settings control how and when ChromeOS devices receive security and feature updates

- **Kiosk settings**: This group contains settings for centrally configuring kiosk computer devices in a secure manner

- **Kiosk power settings**: This group lets you manage the power settings for kiosk devices when they are utilizing AC power or battery power

- **Kiosk accessibility**: This group's settings allow you to configure tools to assist the blind and visually impaired in using ChromeOS devices configured as kiosks

- **User and device reporting**: This group of settings allows you to customize the data collected on users and devices for reporting purposes

- **Display settings**: This group provides a setting for controlling screen resolution and a user's ability to manage it

- **Power and shutdown**: This group's settings determine what power management options will be available and what happens during reboots and shutdowns

- **Virtual machines**: This group allows you to allow or deny the use of Linux VM and ADB

- **Other settings**: This group contains an assortment of miscellaneous hardware configurations

- **Chrome management – partner access**: This group contains only one setting, which allows or denies **Enterprise Mobility Management** (**EMM**) partners to access enrolled devices

- **Imprivata**: This group contains a setting for managing Imprivata identity management software on enrolled devices

After you've explored the many device configuration options you have at your disposal, we have just one more section to tackle – advanced app management.

Exploring advanced application management

Since they play such a critical role in ChromeOS systems and in Google Workspace, apps have already been the topic of discussion several times in this book. In *Chapter 3*, *Exploring Google Apps*, you learned what Google's core app offerings are and what each of the applications is able to do. Then, in *Chapter 9*, *Google Workspace Admin Console*, and earlier in this chapter, you learned where to locate the tools needed to centrally manage applications. Now we'll close out our discussion of apps by examining how to access their global configurations.

To begin configuring the global settings for your Google Workspace apps, you'll need to navigate to the **Apps → Google Workspace** section of the Admin console, as seen in *Figure 10.31*.

Figure 10.31 – Apps → Google Workspace submenu

In this menu, you see each of Google's **Default Apps** listed. By selecting an app name from the list, you are able to launch a configuration page that contains advanced settings unique to that individual application. For example, when you select the **Calendar** app, you'll see a settings screen that looks similar to the following screenshot.

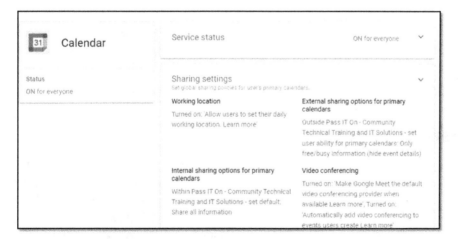

Figure 10.32 – Calendar app settings screen

Although there will be some settings that are universally used across all apps (for example, **Service status**), you can clearly see that other settings, such as **Working location**, are geared specifically toward a time-tracking app such as Calendar. Compare this to an app like **Google Meet**, whose settings can be seen in *Figure 10.33*.

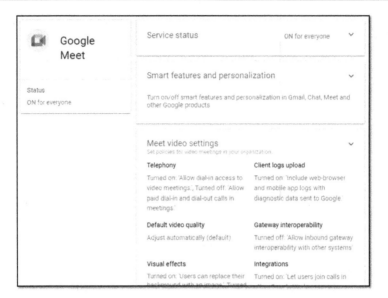

Figure 10.33 – Google Meet settings screen

Here, you can clearly see that advanced management requires an app-by-app approach that goes beyond the scope of our text. However, by exploring each settings page individually and carefully reading the informative descriptions for each setting, you'll have no problem determining which to implement for your environment.

If you want to learn even more about this topic and others related to Google Workspace administration, check out *Google Workspace User Guide: A practical guide to using Google Workspace apps efficiently while integrating them with your data* by *Balaji Iyer* and published by *Packt*.

With that, you've completed the journey to becoming a ChromeOS systems administrator!

Summary

In this chapter, you got the chance to learn about the most used tools and techniques for implementing centralized management in Google Workspace. You went step by step through the process of creating OUs and managing user accounts, explored the tools used to set up groups and target audiences, and got an overview of the many ChromeOS device configurations available to the admin. Finally, you were introduced to the centralized management options for Google Workspace apps and pointed toward additional resources that could be used to master their administration.

Congratulations, you've done it! You have now completed the book and your exploration of the Chrome operating system. You're now fully equipped to administer this powerful, cloud-first operating system at home or on the job. Mastering one of the fastest growing OSs on the market will definitely benefit you as you continue to grow your technical skill set, and with the rising popularity of ChromeOS, there will be no shortage of opportunities to put your new systems administration skills to good use.

Index

Symbols

3-2-1 Rule 99
128-bit AES encryption standard 96
256-bit AES encryption 97

A

Access and data control submenu 193
  API controls page 193
  Client-side encryption 194
  Data classification 194
  Google Cloud session control 194
  Less secure apps screen 195
Account Recovery page 190
  Advanced Protection Program menu 191
  Login challenges page 192
  Password Management page 192
  SSO with SAML applications and SSO
    with third party IdP submenu 193
Account section, Admin console
  Account settings menu 174
  Admin roles menu 174
  Data Migration screen 176
  Domains menu 175

activity controls
  reference link 45
Admin console
  accessing 171-173
  Account section 174
  Apps section 185
  Billing section 177
  Devices submenus, exploring 200
  Directory submenus, exploring 200
  features 170, 171
  Reporting section 178-180
  Rules section 196, 197
  Security section 190
  Storage section 195
advanced application management
  exploring 228-230
advanced device management
  exploring 226-228
advanced user account management
  custom attributes, managing 220-222
  exploring 214-218
  multiple user accounts, adding in
    Google Workspace 218-220
  multiple user accounts, editing in
    Google Workspace 218-220
advanced user management 78
agent 32

allowlisted domains 175

Android Debug Bridge (ADB)

enabling 138-140

Android devices

syncing with 44-46

used, for mobile data connection 37-39

Android Studio 140

app report 180

AppSheet 63

Apps section, Admin console

Additional Google Services page 187

Google Workspace 186, 187

Google Workspace Marketplace

apps menu 189

Overview page 185

Web and mobile apps page 188

Asana 189

B

battery_firmware info command 152

battery_test command 151

Billing section, Admin console

Get More Services page 177

Payment accounts page 178

Subscriptions menu 177

Blogger 187

Bluetooth 40

connecting with 40

enabling, on ChromeOS device 40

C

Calendar 56

casting 42

from desktop 43, 44

from Google Chrome 43

Chat 57, 201

Chrome apps 63

Chromebook 5

Chromebox 5

Chrome browser 89

settings 89-91

Chrome browser-based apps

accessing 63, 64

Chrome Canvas 187

Chromecast

connecting with 41

Chromecast built-in 41

Chrome data in account

reference link 46

ChromeOS 3, 30

cloud focused 4

cost 4

desktop and tools 7, 8

Device menu 13, 14

hardware-based encryption 95, 96

hardware flexibility, with ChromeOS Flex 28

Personalization menu 23, 24

personalizing 7

recovery mode 102, 103

setting up 5-7

software-based encryption 97, 98

system recovery 102

updates, significance 27

upgrade, performing 27, 28

usage 4

USB-based recovery method 103-107

versus competitors 4

ChromeOS, desktop and tools

App Launcher 9-11

shelf 8

status tray 12, 13

ChromeOS, Device menu

Display settings 18-21

Keyboard settings 16-18

Mouse and Touchpad settings 14
Mouse settings 15
Power settings 22, 23
Storage management settings 21, 22
Touchpad settings 15
ChromeOS, Personalization menu
account image, modifying 24, 25
Screen Saver 26, 27
Wallpaper app 25
Chrome Remote Desktop 63, 187
Chrome Shell (Crosh)
accessing 146
scripting 157-160
Chrome Web Store 64, 65, 187
reference link 64
CIA Triad 72
Cipher 94
Ciphertext 95
client-server model 171
cloud-first 3
Comma-Separated Values (CSV) file 181
Condition builder tool 183
Crosh commands, for systems
        administration 146
battery_firmware info command 152
battery_test command 151
enroll_status command 149
exit command 157
free command 152
help_advanced command 147
help command 147
ipaddrs command 149
meminfo command 152
memory_test command 153
modem command 150
network_diag command 155
ping command 150
rollback command 157

route command 156
set_time command 151
storage_test command 154, 155
top command 148
tracepath command 155
uptime command 147
vmc command 157
crosh terminal 36, 121
Crosh Window app 147
Cryptographic Algorithm 94
Cryptographic Key 95

D

data backup strategies, for ChromeOS
backing up, to external storage 101, 102
backing up, to Google Drive 99-101
basics 98, 99
default apps 48, 49
configuration options 51
managing 47
modifying 49, 50
viewing 49, 50
Defense-in-Depth (DiD) 79
desktop
casting from 43, 44
Devices submenu
exploring, in Admin console 200
structure 205-211
Directory submenu
exploring, in Admin console 200
structure 200-204
Discord 128
DisplayPort 41
Docs 57, 187, 201
domain aliases 175
dongle 36
used, for mobile data connection 36, 37
Drive 55, 187

E

encrypted tunnel 32
encryption 32, 94, 95
enrollment 149
Enterprise- and Education-level
    Google Workspace 194
Enterprise- and Education-level
    subscriptions 182
Enterprise Mobility Management
    (EMM) 228
Ethernet 118
    connecting with 30
exit command 157
external display
    connecting 41
external storage devices
    backing up to 101, 102

F

firmware 85
Forms 59, 189
free command 152
freemium plans
    versus premium plan 61
fscrypt encryption 97

G

Gentoo 129
Gmail 187, 189
GNU core utilities (coreutils) 151
GNU Image Manipulation
    Program (GIMP) 131
Google apps 47, 63
Google Assistant 7

Google Calendar 203
Google Chrome
    casting from 42, 43
Google Cloud Platform (GCP) 171
Google Drive 97, 201
    backing up to 99-101
Google Meet 229
Google Play/Google Play Store 65-67, 187
Google Workspace 53
    accessing 53, 54
    Business Plus 162
    Business Standard 162
    Business Starter 162
    Calendar 56, 162
    Chat 57
    Docs 57
    Drive 55, 162
    Enterprise 162
    features 162-169
    Forms 59
    freemium versus premium plans 61
    Gmail 54, 162
    groups and target audiences 222-226
    Jamboard 60, 61
    Keep 59
    Meet 55, 162
    migrating to 197
    multiple user accounts, adding
        and editing 218-220
    overview 162
    personal edition 62
    premium tiers 62, 63
    reference link 197
    shared drives 162
    Sheets 58
    Slides 58
    submenu 229

Google Workspace edition
  Business Plus 62
  Business Standard 62
  Business Starter 62
  Education Fundamentals 63
  Education Plus 63
  Education Standard 63
  Enterprise 63
  Essentials Starter 62
  Workspace of Nonprofits 63
Google Workspace for Enterprise 78
Google Workspace, services
  Chat 225
  Docs 225
  Drive 225
graphical user interface (GUI) 126
groups audiences 222-226
G Suite 53
guest browsing 77, 78

H

HDMI 39, 41
help_advanced command 147
help command 147
Hypertext Transfer Protocol
      Secure (HTTPS) 141

I

initial RAM disk (initrd) 85
instant tethering
  reference link 37
integrated development
      environments (IDEs) 128
Internet Protocol Security (IPSec) 33
ipaddrs command 149

J

Jamboard 60, 61

K

Kanban-style cards 172
kebab menu 112
Keep 59
Kindle Cloud Reader 63
Kiosk & Signage Upgrade 182

L

L2TP/IPSec + pre-shared key 33
L2TP/IPSec + user certificate 33
Layer 2 Tunneling Protocol (L2TP) 33
layered defense 79
Lightweight Directory Access
      Protocol (LDAP) 204
Linux backup/restore
  managing 134, 135
Linux development environment (LDE)
  enabling 126
  features 128-132
  turning on 126-128
Linux distributions
  Debian-based 129
  Red Hat-based 129
Linux permissions
  managing 132-134
Linux storage
  managing 136-138

M

managed domains 175
Manual-touch enrollments 207

master boot record (MBR) 85

Meet 55

meminfo command 152

memory_test command 153

Microsoft's Active Directory 204

mobile broadband network 35

mobile data connections 34

   built-in mobile data 35, 36

   hardware solutions 35

   used, for connecting internet 34, 35

   using, Android device 37-39

   using, dongle 36, 37

modem command 150

N

Network Attached Storage (NAS) device 99

network connectivity issues

   troubleshooting 118-121

network_diag command 155

network interface card (NIC) 30

network speed issues

   troubleshooting options 112-116

notifications 51

   managing 47

Notifications menu

   accessing 52

O

offline files 99

offline syncing 115

One-Time Pad (OTP) 33

Open Authorization (OAuth) 182

OpenID Connect (OIDC) 193

OpenLDAP 204

OpenVPN 33

operating system (OS) 3

   missing, issue 121

Organizational Unit (OU) 202

   organized with 212-214

OS kernel 85

OS recovery 122

Other Google services menu 76

   options 76

OTP card 33

P

parental controls 86

   accessing 86-89

password complexity 6

patching 82

peer-to-peer (P2P) model 170

ping command 150

Plaintext 94

port 141

port forwarding

   implementing 141-143

premium plan

   versus freemium plan 61

premium tiers 62, 63

progressive web app 65

R

recovery mode 102, 103

recovery storage disk (RSD) 86

Reporting section, Admin console 178-180

   Aggregate reports option 181

   Apps Reports 181

   Audit and investigation section 182

   Chrome page 181

   Cost Reports page 181

   Devices menu 181

Email Log Search page 184
Google Workspace Apps Monthly
    Uptime menu 185
Highlights page 180, 181
Manage Reporting Rules submenu 183
Mobile menu 181
User Reports menu 181
rollback command 157
route command 156
routing table 156
Rules section, Admin console 196, 197

S

Sandboxing 85, 97, 136
screen locks 79, 80
secondary user accounts 72, 73
secure tunnel 32
Security Assertion Markup
    Language (SAML) 2.0 193
Security section, Admin console
    Access and data control submenu 193
    Alert center page 190
    Authentication submenu 190
    Overview page 190
Server Certificate Authority (CA) 33
Service-Level Agreements (SLAs) 185
service set identifier (SSID) 31
set_time command 151
Sheets 58
shell script 136
Simple Certificate Enrollment
    Protocol (SCEP) 211
Single Sign-On (SSO) 192
Slides 58
slow-running Chrome device
    troubleshooting 111
Small Office/Home Office (SOHO) 170

Smartsheets 193
Smart TV
    connecting with 41
Software Development Kit (SDK) 194
standalone upgrade 182
status tray 30
Storage section, Admin console 195
    Shared drives using the most
        storage screen 196
    Storage settings 196
    Users using the most storage screen 196
    Workspace storage 195
storage_test command 154, 155
streaming 42
SurveyMonkey 193
syncing 42
    with Android devices 44-46
Sync menu 74
    options 74, 75
system crashing 110
    issue, fixing 110, 111
system freezing 110
    issue, fixing 110, 111
system updates 82
    enabling 83, 84
    performing 83

T

Tableau 193
tab refresh issues
    troubleshooting 116-118
target audiences 222-226
Task Manager 112
Teaching and Learning Upgrade 63
time string command 151
top command 148
tracepath command 155

Transport Layer Security (TLS) 97
Trello 189
Trusted Platform Module (TPM) chip 96
Two-Factor Authentication (2FA) 80-82, 172

U

uptime command 147
USB-based recovery method 103-107
USB-to-Ethernet adapters 30
user account management 72

V

verified boot 85
vim 159
virtual private network (VPN) 30, 32
  ChromeOS device, connecting to 33, 34
vmc command 157
VPN host server 33
VPN token 33

W

WHOIS database 169
Wi-Fi
  connecting with 31, 32
wirelessly tethering 34
Wireless NICs 31
wireless security 31

Z

Zero-touch enrollments 207

Packt.com

Subscribe to our online digital library for full access to over 7,000 books and videos, as well as industry leading tools to help you plan your personal development and advance your career. For more information, please visit our website.

Why subscribe?

- Spend less time learning and more time coding with practical eBooks and Videos from over 4,000 industry professionals

- Improve your learning with Skill Plans built especially for you

- Get a free eBook or video every month

- Fully searchable for easy access to vital information

- Copy and paste, print, and bookmark content

Did you know that Packt offers eBook versions of every book published, with PDF and ePub files available? You can upgrade to the eBook version at packt.com and as a print book customer, you are entitled to a discount on the eBook copy. Get in touch with us at customercare@packtpub.com for more details.

At www.packt.com, you can also read a collection of free technical articles, sign up for a range of free newsletters, and receive exclusive discounts and offers on Packt books and eBooks.

Other Books You May Enjoy

If you enjoyed this book, you may be interested in these other books by Packt:

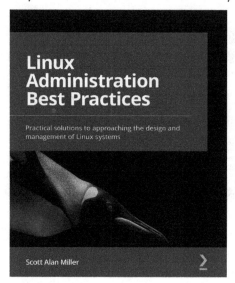

Linux Administration Best Practices

Scott Alan Miller

ISBN: 9781800568792

- Find out how to conceptualize the system administrator role
- Understand the key values of risk assessment in administration
- Apply technical skills to the IT business context
- Discover best practices for working with Linux specific system technologies
- Understand the reasoning behind system administration best practices
- Develop out-of-the-box thinking for everything from reboots to backups to triage
- Prioritize, triage, and plan for disasters and recoveries
- Discover the psychology behind administration duties

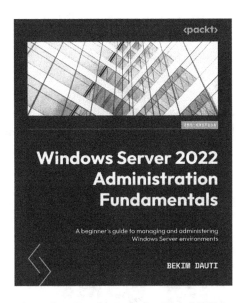

Windows Server 2022 Administration Fundamentals - Third Edition

Bekim Dauti

ISBN: 9781803232157

- Grasp the fundamentals of Windows Server 2022
- Understand how to deploy Windows Server 2022
- Discover Windows Server post-installation tasks
- Add roles to your Windows Server environment
- Apply Windows Server 2022 GPOs to your network
- Delve into virtualization and Hyper-V concepts
- Tune, maintain, update, and troubleshoot Windows Server 2022
- Get familiar with Microsoft's role-based certifications

Packt is searching for authors like you

If you're interested in becoming an author for Packt, please visit `authors.packtpub.com` and apply today. We have worked with thousands of developers and tech professionals, just like you, to help them share their insight with the global tech community. You can make a general application, apply for a specific hot topic that we are recruiting an author for, or submit your own idea.

Share Your Thoughts

Now you've finished *ChromeOS System Administrator's Guide*, we'd love to hear your thoughts! Scan the QR code below to go straight to the Amazon review page for this book and share your feedback or leave a review on the site that you purchased it from.

`https://packt.link/r/1803241055`

Your review is important to us and the tech community and will help us make sure we're delivering excellent quality content.

Download a free PDF copy of this book

Thanks for purchasing this book!

Do you like to read on the go but are unable to carry your print books everywhere? Is your eBook purchase not compatible with the device of your choice?

Don't worry, now with every Packt book you get a DRM-free PDF version of that book at no cost.

Read anywhere, any place, on any device. Search, copy, and paste code from your favorite technical books directly into your application.

The perks don't stop there, you can get exclusive access to discounts, newsletters, and great free content in your inbox daily

Follow these simple steps to get the benefits:

1. Scan the QR code or visit the link below

https://packt.link/free-ebook/9781803241050

2. Submit your proof of purchase

3. That's it! We'll send your free PDF and other benefits to your email directly